DEEP JUNGLE

www.booksattransworld.co.uk

Also by Fred Pearce and available from Eden Project Books

When the Rivers Run Dry

The Last Generation

DEEP JUNGLE

Fred Pearce

eden project books

TRANSWORLD PUBLISHERS
61–63 Uxbridge Road, London W5 5SA
a division of The Random House Group Ltd

RANDOM HOUSE AUSTRALIA (PTY) LTD
20 Alfred Street, Milsons Point, Sydney,
New South Wales 2061, Australia

RANDOM HOUSE NEW ZEALAND LTD
18 Poland Road, Glenfield, Auckland 10, New Zealand

RANDOM HOUSE SOUTH AFRICA (PTY) LTD
Isle of Houghton, Corner of Boundary Road & Carse O'Gowrie,
Houghton 2198, South Africa

First published in 2005 by Eden Project Books
a division of Transworld Publishers
in association with Granada Wild and Thirteen/WNET

This paperback edition published in 2006

Picture Credits: Page 1: Naturepl.com (Bruce Davidson); Page 2: *Top*: Werner Forman Archive (National Museum of Anthropology, Mexico City), *Middle*: Hans Bänziger, *Bottom*: Lou Jost; Page 2–3: Still Pictures (Y.J. Rey-Millet); Page 4: Bridgeman Art Library (Bibliothèque Nationale, Paris); Page 5: *Main*: Naturepl.com (Anup Shah), *Inset*: Fauna & Flora International (Juan Pablo Moreiras); Page 6–7: Fred Pearce; Page 7: *Top*: Still Pictures (Mark Edwards), *Bottom*: Still Pictures (Dominique Halle); Page 8: Chloe Cipoletta

A catalogue record for this book is available from the British Library.
ISBN 9781903919569 (from Jan 07)
ISBN 1903919568

Set in 11.5/15 Minion by
Falcon Oast Graphic Art Ltd

Printed in Great Britain by
Cox & Wyman Ltd, Reading, Berkshire

1 3 5 7 9 10 8 6 4 2

Papers used by Eden Project Books are made from wood grown
in sustainable forests. The manufacturing processes conform to the
environmental regulations of the country of origin.

CONTENTS

Introduction	1
From Darkness to Light	9
Hearts of Darkness	59
Laws of the Jungle	107
Primeval Gardens	151
Shaving the Planet	197
Fruits of the Forest	239
Apes, Ancestors and a New Human Dimension	281
Further Reading	306
Index	309
Acknowledgements	313

INTRODUCTION

It is the most extreme, the most complex place on Earth – perhaps in the whole universe. But is it a green hell or a green heaven? A place of exquisite beauty or unimaginable horror? The cradle of humanity or our most alien terrain? Something to be destroyed or treasured? We seem always to have had an ambivalent relationship with this most brilliant manifestation of our planet's natural wealth.

Once the jungle was an impenetrable, alien world of myths and torments that still, somehow, held unaccountable, unimaginable delights. The Spanish conquistadors went again and again in search of El Dorado, the jungle city of gold. The nineteenth-century German explorer Adolphus Frederick found a 'deadening, soul-killing forest, oppressive in its monstrous hugeness and density' – but he still went back. Henry Morton Stanley, Welsh orphan and explorer, renowned for his search in Africa for the Scottish missionary David Livingstone, said it was 'a murderous world' and its people 'filthy, vulturous ghouls'. He went back, too.

But many explorers found Eden. The French Romantic

painter Henri Rousseau portrayed it best in his painting *The Dream*, in which a naked woman reclines on a couch in a moonlit jungle, surrounded by plants and animals. This was a wild paradise, home of the noble savage and – as another German explorer, Alexander von Humboldt, put it – of an 'infant society that enjoyed pure and perpetual felicity'.

Europeans have been perpetual invaders of the forest. We have gone for gold and for power, and to find an empty canvas for our imagination: 'My last chance to be a boy . . . in the last great wilderness of Earth', said retired US President Theodore Roosevelt as he headed for the Amazon in 1913. And we have gone for intoxication: in the Sixties, students 'turned on, tuned in and dropped out' under the influence of drugs found in the jungle by crusty professors a generation before. In the 1980s, the German film-maker Werner Herzog, in his *Burden of Dreams*, saw 'overwhelming misery and overwhelming fornication, overwhelming growth and overwhelming lack of order'.

Today we have renamed the jungle the rainforest and given it to science. We revere it as the crucible of evolution, the birthplace of humanity and the heart and lungs of the planet. We go there searching for the origins of evolution and for drugs to fight AIDS. The new El Dorado is biological. For the venerated American rainforest biologist Edward Wilson, the rainforest is 'timeless, immutable'. A British compatriot calls it 'the finest celebration of nature that has ever graced the face of the planet'.

This book is an attempt to get to grips with the romance

and the reality of the rainforest. It tries to delve deeper into the jungle, its byways, its history, its laws and its cacophony of wildlife. And it offers a new twenty-first-century perspective that combines a proper sense of wonder and awe with myth-busting scepticism about some of the romantic nostrums that still fill our visions of the world's greatest and least-known ecosystems.

We go to the heart of the jungle with the scientists who can claim to know it best. They spend their days hacking through undergrowth or flying through the canopy like latterday Tarzans, then dine out on jungle bugs and bushmeat and spend their nights sharing mind-blowing hallucinogenic drugs with shamans – all in the interests of science, of course. But they are serious researchers and there is no hiding place from these Big Brothers of the jungle. They measure and count and track and photograph and listen for everything. They shine ultra-violet light through the canopy, probe the jungle floor with x-rays and radar, and watch it all from satellites circling overhead. They listen for the vibrations made by soft-footed mammals, for sweet-songed birds and for the ultrasonic screeches of bats. They dig up the forest floor, sift through orang-utan dung and sniff for the forest's breath high in the canopy. They chase tiny seeds through the jungle and put genetic markers on the wasps that carry them. They strap micro-cameras on to the backs of ant troops and track the night-time manoeuvres of mammalian foot soldiers with heat-sensitive metal lenses and military night-scopes. They fly remote-controlled cameras through the canopy, collar elephants and paint mice with fluorescent powder in order to watch their multi-coloured flight through the branches.

But still they stand in awe at the wonder of the place – at the hour-long displays of birds of paradise high in the trees of New Guinea; at the scary, quasi-human politics and warfare and sex of the great apes; at the vast, hidden ecosystems in the canopy, where snakes and earthworms live and die without ever coming to earth; at the gigantic dark 'cathedrals' beneath the canopy, and the sheer blood-sucking, flesh-eating, mind-boggling, poisonous menace of evolution on speed.

We shall also meet bushmeat hunters and seed smugglers and men still seeking El Dorado; tear-drinking insects and new species of elephant and tiger; the sharpest teeth in the jungle; botanic monsters that strangle everything in their path – and the moth with an 11-inch tongue. And I hope we shall find answers to some baffling questions.

Just how virgin are the virgin rainforests? Can it be true that the mythical Amazon city of El Dorado was there all along, hiding in the foliage – and what links it to African ramparts buried in the bush and bigger than the Great Wall of China? Why might the banana and the peanut both be doomed? Was Darwin right about the 'survival of the fittest' – or are new theories about the 'survival of the weakest' nearer the mark? Is the jungle a closet socialist? Is it possible that oil prospectors and cocoa growers and even slash-and-burn farmers could be good news for forest conservation? Who owns tribal knowledge? And just what should jungle hunters have for lunch if we deny them bushmeat?

Somewhere along the way, the forbidding, alien 'jungle' was renamed the inviting, fecund 'rainforest'. The new El Dorado is biological, but also spiritual. And yet, for all our new-found infatuation, we still fear the forest and conjure up myths about it. We just cannot quite work the place out. Perhaps there is something truly primeval here, a folk memory of our origins as a species that still spooks us. Perhaps the forest is where our hearts and imaginations truly reside, and where we go to learn about ourselves.

From the golden city of El Dorado to modern environmentalism, the jungle seems always to be at the heart of things, reflecting our desires and fears. It still shocks and appals us; still amazes and baffles. Maybe that is because we still know so little about it. Maybe it is because in it, we see ourselves. Or maybe that is just another myth.

YELLOW RAIN

One of the most notorious modern myths of the rainforest succeeded in combining the mystery of the jungle with the paranoia of the Cold War. In September 1981, Ronald Reagan's Secretary of State, Alexander Haig, announced to an incredulous press conference that 'for some time now, the international community have been alarmed by continuing reports that the Soviet Union and its allies have been using lethal chemical weapons in Laos, Kampuchea [Cambodia] and Afghanistan . . . We now have physical evidence from southeast Asia which has been analysed and found to contain abnormally high levels of three potent mycotoxins – poisonous substances not indigenous to the region and which are highly toxic to man and animals.'

He had, in fact, a couple of leaves from remote Cambodian rainforest close to the border with Laos that carried on their surface tiny amounts of natural poisons called trichothecenes. Some in the Pentagon believed the Russians were manufacturing these chemicals as weapons, and a hue and cry ensued. Here on these leaves was proof that Russia was waging biological warfare in the jungle. The revelation excited defence analysts and elicited long articles in the Wall Street Journal analysing the threat. The Reagan administration accused Russia of violating arms control agreements, and claims and counterclaims followed for years. Then, in 1987, Harvard biochemist Matthew Meselson came back from the jungle with samples of bee faeces and proved to the satisfaction of all but a few – who can still be found busily making their case on the Internet – that the yellow rain was non-toxic bee faeces: perfectly natural bee shit produced from time to time by swarms of giant honey-bees in the forest, apparently in response to heat stress. The spooks had been spooked by the rainforest.

And the story had a splendid sequel. In 2002, a nervous India was gripped by fears that it, too, had come under attack from chemical weapons. For two days, green rain fell on the town of Sangrampur close to the mangrove forests of West Bengal, spattering clothes and buildings. Villagers rushed to temples to pray, and rumours spread that the rain was contaminated with chemical warfare agents. When the state's chief pollution scientist arrived from Calcutta, he reported that the mysterious yellow-green droplets were in fact bee faeces, coloured by pollen from local mangoes and coconuts. He said the green rain was caused by a mass migration of a swarm of giant Asian honey-bees, whose propensity for producing 'golden showers' is well known to bee biologists. Panic over . . . until the next time.

FROM DARKNESS TO LIGHT

From the day Christopher Columbus set foot in the Americas, and Portugal's Henry the Navigator set his sailors ashore in West Africa, Europeans went to conquer the jungles with bravado, lust and fear. They regarded these steaming, fetid, alien lands as both places 'of the utmost dread' and 'the entrance to paradise'. First they went in search of El Dorado, a mysterious jungle city of gold, and of Christian kings like Prester John. But they were seduced, too, by weird hallucinogenic plants that blew their minds, by the sheer biological wonder of this 'inexhaustible treasure trove' and, finally, by its greatest secrets – the real El Dorado, the mysteries of evolution.

IN SEARCH OF EL DORADO

Christopher Columbus, in his log dated 28 October 1492, reported stepping ashore amid the forests of Cuba. 'I have never seen anything so beautiful,' he wrote. 'The country is full of trees, beautiful and green and different from ours, each with flowers and its own kind of fruit. There are many birds of all sizes that sing very sweetly, and there are many palms.' It was an idyllic scene – but not what he had come for. From the day they first set foot in the Americas, the Spanish were seeking gold. Columbus had persuaded his Royal patrons back home that his journey, to what he presumed to be Asia, would deliver the contents of the 'gold mines of Japan'. So, rather than lazing under the palm trees, bird spotting or fraternizing with the natives, Columbus got down to business: 'The search for gold began the day after our arrival.'

As Columbus, his fellows and successors spread out across the New World, conquering the Aztecs in Mexico and the Incas in Peru, and scattering the numerous tribes of the Amazon, there was a manic fervour about their search for gold that combined the mercenary and the spiritual. Columbus himself was clear about his dual motivation: 'Whoever possesses gold can acquire all that he desires . . . with gold he can gain entrance to paradise.' As one of the most successful conquistadors, the cruel and handsome Hernando Cortez, put it after sacking the Aztec capital: 'We have a sickness of the heart for which there is only one cure – gold.'

It wasn't hard to find gold in the halls of the mountain peoples that the Spaniards swiftly conquered. Civilizations like the Chimus, Moches and Incas of Peru, the Chibcha of Colombia and the Aztecs of Mexico were all avid workers of gold. But even as they sacked these kingdoms, the conquerors appeared dissatisfied with their plunder. They began to believe that there was more, much more, gold buried deep in the continent's darkest corners, in the forests of the Amazon lowlands. It is hard to understand where this belief came from, but rainforests and myths about gold seem always to have gone together. Gold, the most valued but least attainable commodity throughout the Middle Ages, seemed to Europeans to be buried, of necessity, in the most inhospitable and impenetrable places – the forests of distant and alien lands.

Many believed that the 'perfect metal' was an organic part of the forest and grew in the jungles like a living organism; and experiences in the Americas did not dent that belief. The late-sixteenth-century Jesuit naturalist Jose de Acosta, who wrote an early natural history of Peru, argued that 'minerals seem to grow like plants . . . they emerge from the bowels of the earth as a result of the virtues and efficiency of the sun and other plants, so that over a long time they continue to grow and almost propagate.' To put a missionary spin on things, the priests argued that God put gold among the primitive forest people to encourage Christians to seek them out and convert them. As one priest put it: 'A father with an ugly daughter gives her a large dowry to marry her; and this is what God did with that difficult land, giving it much wealth in mines so that by this means he would find someone who wanted it.'

* * *

But if the forest was the door to paradise, it was also a hell on Earth. Its invaders went in fear of demons, both animal and human, real and imagined. By day, as they slashed their way through the jungle, the conquistadors faced snakes and jaguars, natives with poisoned arrows and mosquitoes carrying killer diseases. And as they set up nervous camp each night, they swapped stories about serpents that could swallow a man whole, and of a tribe whose members had no heads, but eyes in their shoulders and mouths on their breasts. They muttered, too, about a tribe of warrior women who procured and discarded slave men at will. These Amazons were 'very white and tall, go about naked with their bows and arrows in their hands, doing as much fighting as ten Indian men', wrote Gaspar de Carvajal, a Dominican missionary who went on one of the most famous expeditions into the Amazon jungle in 1540 with his captain, Francisco de Orellana.

The myth of a tribe of female warrior Amazons was not

new. Like the stories of gold in the jungle, the invaders had brought it with them from Europe. Indeed, the Amazons had a history dating to Greek times: for a long time they were believed to live near the Black Sea, on the borders of Europe and Asia; but ever since de Orellana and de Carvajal claimed to find them in the Brazilian rainforest, and named after them the river down which they journeyed, the Amazons have been relocated in the world's imaginings. Right into the nineteenth century, serious scientists continued to argue in favour of their existence somewhere in the jungles of South America.

But even more than the mythology of the Amazons, the conquistadors came, most of all, to believe in the story of El Dorado, a city of gold deep in the Amazon jungle. Nobody knows quite how the myth got going: some say it had its origins in the traditions of the Chibcha people in the mountains of modern-day Colombia, who said the mountains had once been ruled by an Indian chief who was so wealthy that every morning he got his servants to cover him in gold dust, which he washed off in a lake. However it arose, the myth of El Dorado soon lost all its geographical bearings as conquistadors went up the Orinoco and down the Amazon, over the Andes and into the forests on the continent's Pacific shores, seeking out the fabled king and his city.

If they had a direction at all, it was generally to the continent's great interior, into the largest jungle on Earth. There, said one guide for prospective conquistadors, seekers of gold 'should try to enter inland as far as they can – for they cannot fail to find great secrets and riches'. From Venezuela and

Peru, Panama and the Guyanas, most seekers of El Dorado make a beeline for the heart of the continent – for the jungle fastness where gold must be there for the taking. But where in the jungle? The Amazon jungle is a big place. Some said it was in the west, twelve days' march from Quito; others that it was somewhere in the headwaters of the Orinoco, in the north of the rainforest region; still others mapped it further east, close to the home of the Amazons, who ruled a land somewhere deep in the rainforest that was 'very rich in gold'. But in their fervour to join the gold rush, it scarcely seemed to matter: the lure and the chase were the thing.

* * *

Who were these conquistadors seeking El Dorado in the jungle? One of the first was Diego de Ordas. He had already become one of the richest men in the world after sending home huge amounts of loot gathered while he was Cortez's captain during the sacking of the Aztec capital of Tenochtitlan in Mexico in 1520. But he wanted more. He got himself appointed governor of Maranon, the first name given by the Spaniards to the Amazon, and set about putting together an expedition to go looking for gold near the source of the Orinoco. His watery journey up South America's second greatest river and through the jungle became the prototype of all future searches for El Dorado, a journey through a dreamland alternating between heaven and hell, in search of a paradise of riches.

Disease and fever were everywhere, he recorded. 'This

region was so terrible and the vapours so corrupt and heavy that if someone was bitten by a vampire [bat] or got a small cut he immediately became cancerous. Men from one day to the next had their entire feet consumed by cancer. They were dying one by one.' There were jaguars in the jungle, cayman and anacondas in the river, and the air was filled with mosquitoes and black flies. Their bodies became infested with worms that buried themselves beneath the skin and grew till they were several inches long.

Things looked up when the crew met some Indians from the Aruak tribe. The expedition reported, with perhaps understandable descriptive zeal, that the women were naked except for 'a rag in front of their private parts, which is loose. When they sway, or in the wind, everything is revealed.' And there was more, according to the great English historian of the period John Hemming: 'The tribe provided an attractive woman to sleep with any stranger; and when he left, the woman was free to go with him.' But it was too good to last – the conquistadors were not so easily seduced away from their warrior yearnings. When the Aruak people innocently dropped by at the conquistadors' camp to look at a collection of pigs they had brought for food, Ordas decided that they were planning to slaughter the pigs, so he and his men slaughtered the Indians instead. From then on, the Indian women were forcibly taken rather than granted as gifts.

As tales of mythical gold spread, Ordas was soon joined in the jungle by motley bands of hopeful soldiers, adventurers and even civil servants. One of the latter was the Royal treasurer Jeronimo Dortal, who had had enough of cataloguing the booty from the conquests back home in Europe, and sought to make his own fortune. He set out to find a new route south to the Orinoco headwaters that avoided the treacherous and fever-ridden delta region, but he ended up alone after suffering a mutiny.

Outstanding among these journeys, however, was the expedition organized by Gonzalo Pizarro, brother of Francisco, the conqueror of Peru. He and his companion and second-in-command, Francisco de Orellana, set out east from Peru in February 1541. It was a vast expedition: Pizarro and de Orellana took with them 4,000 Indian porters, 200 horses, 3,000 pigs, as well as llamas and packs of hunting dogs that were trained, it was said, to attack Indians. It was a brutal and ill-fated affair. Pizarro, in particular, tortured his porters for not telling him what lay ahead, when these peasants from the Peruvian mountains knew no more than he what the jungle contained. Likewise, he used his dogs to tyrannize every forest native who claimed not to know the whereabouts of El Dorado. 'They must be liars,' he said.

Within a month, three thousand of Pizarro's Indian bearers had died or deserted, taking the pigs with them. So had many of his Spaniards, for whom even the lure of gold was not enough. Without hunters or pigs, they were also starving. De Orellana took a small contingent downstream on a tributary of the Amazon to find food, but he was unable to return because of the strong current. Pizarro failed to follow, so de Orellana sent messengers back overland. They discovered that Pizarro had hightailed it back to Quito, which he eventually reached with just eighty men left.

Perhaps relieved to be rid of his impulsive master, de Orellana decided not to follow. Instead, he headed on downstream into the Rio Negro, which he named, and eventually to the main stream of the Amazon.

De Orellana's journey continued for months. Along the way he made many contacts with Indian communities and had several skirmishes. In June, four months after he and Pizarro had set out from Peru, he fought off an attack from warrior women. Thus he initiated, or at any rate enhanced, the story of the lost female tribe of Amazons from Asia Minor who were, for reasons never explained, now alive and well and living in South America. In late August, he reached the open Atlantic Ocean, becoming the first European to travel the width of the Amazon basin. He returned for a while to Spain where he was fêted for his voyage – and told stories of seeing big roads in the jungle that he believed must lead to a fabulous kingdom. Five years later, de Orellana returned to the jungle in a second attempt to find the city of El Dorado or, failing that, to construct a city of his own. But he achieved neither and died somewhere on the Amazon.

* * *

The invasion of the Amazon in search of El Dorado was rapidly becoming a major undertaking. As Hemming puts it: 'For sheer endurance, mileage walked and tribes, hills and rivers "discovered", these exploits far exceed the famous travels of the nineteenth-century African explorers.' Strange, when all the gold so far found had been in the drier and cooler highlands of the Peruvian and Ecuadorian Andes and the Mexican highlands; but each failure, each new outbreak of fever and attack by jaguar or insect, simply relocated the fabled city somewhere further east and triggered yet more journeys.

Adventurers from other nations joined the hunt. In 1541, as Pizarro and de Orellana squabbled in the headwaters of the Amazon, a German named Philip von Hutten led a party up the Orinoco in search of a reputedly fabulously wealthy tribe called the Omaguas. He found a heavily populated area, but no golden city. The Portuguese, meanwhile, went searching for El Dorado in northeast Brazil; and Walter Raleigh, the sometime favourite of Queen Elizabeth of England, twice sailed up the Orinoco in search of 'that great and golden city', which he called Manoa. He eventually wrote a best-selling book, *The Discovery of the Empyre of Guiana*, that did much to popularize the legend of El Dorado across Europe.

Raleigh wrote of the forest through which he travelled as 'a country that hath yet her maidenhead, never sacked, turned nor wrought, the face of the earth hath not been torn, nor the virtue and salt of the soil spent'. This virginal paradise held terrors: his hundred-strong crew huddled together in their small boats, attacked by natives and alligators. But it seems to have been the smell of his own men that got to Raleigh most: 'What with the victuals being mostly fish, and the wet clothes of so many men thrust together, and the heat of the sun, I will undertake there was never any prison in England that could be found more unsavoury and loathsome.' Having once been locked up in the Tower of London by his Queen for bedding one of her chambermaids, Raleigh knew all about prisons.

Thanks to their earlier sacking of mountain empires, many of the conquistadors became some of the richest people in Europe – like the billionaires of a modern dot-com boom, which, not coincidentally, has itself often been called a gold rush; and the story of the search for El Dorado spoke to the heart of a new yearning in Europe to explore and conquer the world in pursuit of personal fortune. According to some commentators, El Dorado was the first great symbol of what became the American dream – 'a myth of self-creation, personal escape and social transformation', as writer Mark Cocker put it. But while we can recognize that yearning, it is hard today to understand how the myth of an El Dorado in the jungle persisted against all the evidence. Somehow, the failure of each expedition seemed simply to raise the stakes and embellish the original tale. Was it all madness? Well, perhaps so. But, as we shall see later, all those stories of great jungle cities were not necessarily myths. They may have been there, though they disappeared into the jungle almost whenever and wherever the conquistadors went.

The conquistadors meanwhile went looking for El Dorado in rainforests even beyond South America. In 1526, Alvara de Saavedra, after loading up with cloves in the Moluccas islands in the East Indies, on the other side of the Pacific Ocean, reported that he had landed in Isla del Oro, the Land of Gold. This was the giant forest-covered island of New Guinea. It is not clear whether he found gold, but he clearly expected to. Subsequent explorers moved on to the equally forested Philippines, from whence reports soon emerged of an El Dorado where communities lived by mining gold. Here, at any rate, there was gold, which the Spaniards assiduously looted, much as they had in the Andes. They took it back to Acapulco in Mexico, before taking it on to Spain in a fleet of huge new galleons built for the purpose. But they failed to find the gold mines, which the natives of the Philippines kept hidden from them.

Throughout the sixteenth century and on, the rainforests of the tropical regions of the world were almost literally mapped in gold. Nothing else there mattered and, since gold was rarely found and nothing else was properly observed, we are left with a mythological landscape, where paradise and hell almost become one. Almost unnoticed at the time, millions of natives died as the dreamsmiths rampaged through the rainforests. Most died not from direct attack by the invaders, but from the diseases they brought, against which the native populations had no defence. Smallpox arrived in the New World in 1518, measles in 1530, typhus in 1546 and flu in 1558.

By comparison, the invaders got off rather more lightly in this clash of disease cultures. Only one major disease appears to have made it back to the Old World with the returning invaders: syphilis, which burst across the continent after French troops invaded the port of Naples in 1495. The disease claimed an estimated 10 million victims in the next fifteen years. This is a lot – but probably a small return for what happened unnoticed in the New World.

ERROL AND SHARKEYE; CITY Z AND BRE-X

The conquistadors never found El Dorado, but the dream continued. Even in the nineteenth century, after three centuries of failed attempts to locate it, maps still appeared that marked the 'golden lake' of El Dorado somewhere between the Orinoco and the Amazon. And long into the twentieth century, a great many jungle expeditions were being carried out, at least in part, as a search for gold. In 1922, in the interior of the vast, unexplored island of New Guinea – Alvara's Isla del Oro – legendary Australian loner and explorer William 'Sharkeye' Park found gold while panning among the trees on the remote, far side of the central highlands. Sharkeye's find started one of the more unlikely gold rushes, in which fortunes were made in great secrecy by adventurers brave enough to go there, all against a backdrop of head-hunting and cannibalism.

But for every successful gold-digger in the jungle, there were many failures, fraudsters, dreamers and probably many more adventurers who just passed through. One of these last was Errol Flynn. Years before he found fame and fortune as a swashbuckling movie heart-throb, Flynn spent some time in New Guinea, where he found a little gold, traded some plumes plucked from birds of paradise, received a wound from a poisoned dart, was gaoled for killing a Papuan, and left behind him 'much ill will and a swag of debts', according to local accounts.

Among the notable dreamers was Colonel Percy Harrison Fawcett, an eccentric British army surveyor who, in 1910, helped Bolivia mark its forest border with Brazil. During that expedition, Fawcett famously faced down local Amazon Indians, who were aiming their poisoned darts at him from across a river, by playing an accordion and organizing an impromptu dance among his followers until, as one witness in his party put it, 'the ice was broken'. He might have retired home to England after that but, in 1925, fifteen years older but apparently less wise, Fawcett walked into the Amazon forest with his teenage son, clutching on old map with 'City Z' marked on it, in a personal quest for El Dorado. Some say he had dreams of setting up in the jungle a commune dedicated to theosophy, a vision of divine nature based on Buddhism that was popular at the time. The 58-year-old was also reputedly an aficionado of the works of Rider Haggard and of Sir Arthur Conan Doyle's 1912 epic adventure of dinosaurs and ape-men in the Amazon jungle, *The Lost World*. Maybe the two worlds got mixed up: before he went, Fawcett wrote how 'the forest in these solitudes is always full of voices, the soft whisperings of those who came before'. Following these voices, he and his son headed into the forest and were never seen again.

That should have been the end of the Fawcett story: there are many ways to die undetected in the jungle that do not require outlandish theories. And yet, when it comes to the dreamscapes of the rainforests and the myth-making of El Dorado, nothing is ever allowed to be simple. Myth has piled upon myth and, since Fawcett disappeared, no fewer than thirteen separate expeditions over eight decades have gone into the jungle to try and find him, his remains or his 'City Z' – all, predictably enough, without success. In 2004, one British newspaper was recycling the theories of a TV director about how Fawcett was 'perhaps lured by a native she-god or spirit guide whose beautiful image . . . appears only to the Fawcett family and to those who try to track the expedition's path'.

Raleigh, Conan Doyle and the twin lures of gold and the jungle also inspired an American bush pilot and aeronautics wizard from Missouri called Jimmy Angel. He spent much of the 1920s and 1930s in the headwaters of the Orinoco, looking for a 'river of gold' in 'a land where the plesiosaurs roam'. The story told is that he once found gold on a mountaintop, with the help of an Alaskan prospector, James McCracken, whom he had met in a bar in Panama City. But McCracken died soon afterwards and Angel could never find his way back to the right mountaintop. Angel eventually crashed to his death in 1956, still looking for his El Dorado, but on his journeys over the forest he did spot the world's tallest waterfall glinting among the trees in a remote region of Venezuela. That, at least, was for real and his ashes were scattered amid the falling water.

Finally, during the 1990s, the forests of Borneo became the scene of the biggest business fraud of all time – a fraud over gold. An obscure Canadian mining company called Bre-X rose to be one of the most valuable mining companies in the world. Stocks soared on the basis of claims from its small team of explorers that they had found the world's richest gold mine at Busang, a tiny airstrip in the middle of the jungle, 1,500 kilometres from Jakarta. Thousands of investors, including the ruling Indonesian Suharto family, bought into the company. At their peak in 1997, Bre-X shares were worth more than 4 billion dollars, and the entire gold hoard was valued at 25 billion dollars. El Dorado was back in the headlines, this time on the business pages.

Then doubts crept in: questions were asked about the authenticity of the ore samples being assembled and analysed at a remote jungle laboratory. Sceptics suggested that the samples might have been adulterated – not a difficult task since there was no outside supervision and local streams being panned by tribesmen routinely contained small amounts of gold that could be used to 'salt' the samples. Then, on his way to answer the accusations, the chief field geologist Michael de Guzman – a karaoke-loving Filipino with a pronounced limp and four wives – leapt from his helicopter into the jungle. By the time they found his body, half-eaten by forest animals, a few days later, the mine had been deemed worthless and the company's value had crashed to nothing. Like a latter-day El Dorado, the Busang gold mine had turned out to be a forest mirage created by the lure of impossible dreams, sustained by criminality and undetected because it happened in a place where nobody in their right mind wanted to go.

PRESTER JOHN'S LEGACY

Africa south of the Sahara has always been Europe's 'dark continent', and a land of mythical beasts. In the chronicles of the Middle Ages, its forests were peopled variously by half-humans with one leg, three faces and the heads of lions; by one-eyed people who used their feet like monkeys to cover their heads; and by birds that could carry elephants into the air. Such misunderstandings arose in part because Europe was cut off from the object of its fantasies by hostile Arabs, who dominated trade routes and, in effect, besieged Europeans on their own continent. Africa was the world beyond, a jungle land 'of the utmost dread' surrounded by a 'sea of darkness'. And yet, all was not darkness. Rather as the Amazon was later held to harbour a city of gold, it was widely thought that parts of Africa were ruled by a Christian king called Prester John, whose kingdom was a promised land amid the horror, possibly under Arab siege. Europeans wanted both to enter this promised land and to save it.

Prester John's origins went back to the eleventh century, his story probably invented by clerical propagandists intent on launching Christian crusades against the Arab world. His kingdom, which the crusaders sought to relieve from unknown Islamic terrors, became a repository of all kinds of myths – of unicorns that killed lions, of snakes with two heads and horns like rams, of ants as big as foxes and of the 'fount of youth'. Its geography, like that of El Dorado, was rather uncertain: it was variously in India, or Sudan, or beyond Persia, before moving, by the fourteenth century, to the African rainforests beyond the Mountains of the Moon, which would put it in the Congo basin.

Europe's geography began to buck up only with the arrival of Portugal's great explorer king, Henry the Navigator. Gradually the Arab siege was being broken and Henry sent ships tentatively down the coast of West Africa, past Morocco and the shores of the Sahara to drop anchor in the swampy malarial lagoons and forested river inlets of equatorial West Africa. But while the geography was improving, the mythology was unreconstructed. Henry's navigators were in search of the 'rivers of gold' foretold in the stories of Prester John – and they found them. In January 1472, a Lisbon merchant, Fernao Gomes, anchored off the estuary of the river Pra near modern Ghana, where villagers showed him gold dust from the Ashanti mines. He bought the gold in exchange for cloth, but what the Ashanti really wanted was slaves to work the mines. So, within a decade, Portuguese traders were cruising the West African shores, buying thousands of slaves a year from local African tribes and selling them to the Ashanti in return for gold. In 1482, still a full decade before Columbus fetched up in the Caribbean, Diogo Cao sailed beyond Ghana, south across the equator and into a mass of water that 'for the space of 20 leagues preserves its fresh water unbroken by the briney billows'. He had found the mouth of the Congo, the world's second biggest river and entrance to the world's second biggest rainforest. There, Cao erected a limestone pillar and, within a few years, there was a slave port that exported more than five thousand Africans a year to Ashanti.

Just sixteen years later, another Portuguese explorer, Vasco da Gama, rounded the southern tip of the continent. He reported seeing dhows 'laden with gold' sailing out of the forests of East Africa, through the Zambezi delta and into the Indian Ocean. But that was as far as things went, for, having skirted its coastal extremities, Europeans rarely went into the forest interior. They knew something of the plains of southern Africa, where the equable climate had attracted first the Dutch and later the British; they learnt a little of the great lakes of eastern Africa; but for centuries they knew virtually nothing of the great Central African jungle region of the Congo basin, an area the size of western Europe. If they heard tales of its leaders – of monarchs like ManiKongo, who sat on a throne inlaid with ivory, wielded a zebra-tail whip and wore animal heads on his belt – they were not interested in learning more. Prester John could still have been living there, for all they knew.

Perhaps the rainforest dreams of Europeans were sated in the Americas, where gold came only from conquest. Maybe they were happy to buy the gold and ivory, and slaves were easily plundered or bought for trinkets from coastal tribes trading with the interior. As the religious myth of Prester John faded from consciousness, it was never replaced in Africa by the secular dream of an El Dorado or the sexual fantasies of the Amazons. It largely remained the dark continent.

ELIXIRS AND THE BOTANICAL EL DORADO

Jungle bounty, from the earliest explorations, came in the form of plants as well as metal. From the Pizarro brothers onwards, some of the earliest voyages by the conquistadors had, as a subsidiary aim to the discovery of El Dorado, the penetration of the Amazon's supposed cinnamon forests in mind. This was part of the continued confusion about whether the New World was, as Columbus had supposed, the easternmost part of Asia. The cinnamon spices that had reached Europe from the Orient for centuries were grown far away in the East Indies. There was no cinnamon in the Amazon, but the conquistadors found plenty more plants to beguile them. Making their way on to the highlands of Mexico, they swiftly spotted the value that the Aztecs placed on a bean growing in pods on an orchid vine in the coastal rainforests. This was the vanilla bean, which the Aztecs used to flavour something called chocolatl that they made from local cacao beans – another forest product.

When Hernando Cortez visited the court of Montezuma, he noted before destroying the place that the king 'took no other beverage than the chocolatl, a potation of chocolate, flavoured with vanilla and spices, and so prepared as to be reduced to a froth of the consistency of honey, which gradually dissolved in the mouth and was taken cold'. The fact that Montezuma consumed his chocolatl in goblets before entering his harem suggested that it was an aphrodisiac, though there is some lack of clarity about whether it was the vanilla or the cacao beans that was the most potent element in the brew. The Aztecs probably cared little – they mixed their drugs promiscuously. It was claimed that at his coronation in 1502, two decades before Cortez spoilt the party, Montezuma gave his guests chocolatl mixed with the local hallucinogenic mushroom, Pscilocybe, as a special treat.

The Aztecs were botanical masters. The Spanish physician Francisco Hernandez, who explored Mexico in the 1570s, reported back to his king, Philip II, on what had clearly been a sophisticated and highly organized use of native forest plants for medicines and flavourings. Many of the plants had been cultivated in the extensive botanical gardens of Tenochtitlan prior to its sacking. Most of the information gathered by Hernandez fell by the wayside in an era when gold was the only true quest. But some finds made it back to Europe: by 1602, Queen Elizabeth's apothecary in London had received samples of vanilla, to which the Queen was partial. But efforts to cultivate vanilla failed, as they have in most places outside Central America. The vanilla vine requires a specific local pollinator, the tiny melipone bee, in order to reproduce. Europe lost interest and it was another century before a regular trade in the bean began, after which the French in particular took to flavouring their chocolate with vanilla, much in the manner of the Aztecs.

The mountain peoples of the Andes, like the Aztecs in Mexico, were assiduous collectors of useful plants from the surrounding rainforests. A favourite was the coca plant, which is, of course, the source of cocaine, and today is probably the most profitable, albeit illegal, crop in South America. The plant originated in the jungles of the central Amazon, which is where it was probably first cultivated. The locals sucked the leaves as a mild stimulant and burned them during ceremonies to make smoke and induce trances. Peruvian mummies of 1,700 years ago were buried with bags of coca leaves to take to the afterlife with them.

The conquistadors who invaded Cuzco in modern-day Peru in 1533 noted that 'the herb is so nutritious and invigorating that the Indians labour whole days without anything else'. Their successors noted this too, and handed out coca leaves to Indian slaves before driving them to work the silver mines. Back in Europe, coca took a while to catch on – several centuries, in fact. It became popular only in the 1860s, after German chemist Albert Niemann isolated the alkaloid cocaine in the plant and a Corsican entrepreneur called Angelo Mariani added the magic ingredient to a concoction he called 'Vin Mariani'. The new, interestingly fortified wine was celebrated by everyone from Buffalo Bill to Emile Zola, Queen Victoria to the cardinals of the Vatican and Jules Verne to Arthur Conan Doyle. Soon afterwards, John Pemberton, a pharmacist from Atlanta, Georgia, came up with the most lucrative coca product of all when he substituted sugar for wine and created the first recipes for Coca-Cola. He sold his drug-spiked beverage as 'the temperance drink'. Coca was eventually dropped from the product and lovers of its stimulation turned to the refined alkaloid – Sigmund Freud was an early aficionado. But coca's antisocial traits were seriously unleashed only in the late twentieth century, when its addictive qualities were amplified by smoking the pure alkaloid in crack.

British botanists meanwhile had gone to the jungles of the New World in search of their own, less mind-altering elixirs. Reputedly, the first pineapple to be raised in Europe was grown in southwest London and fed to King Charles II, who posed with it for an official portrait. A decade later, in 1689, the Duke of Albemarle's personal physician, Hans Sloane, brought back a huge collection of forest goods from Jamaica. His baggage included eight hundred plants, among them cocoa, a 7-foot-long yellow snake from the forests and the embalmed remains of his recently deceased employer, who had been governor of the island.

Having seen Jamaicans prepare a drinking chocolate from cocoa, honey and pepper, Sloane came up with the idea of mixing it with milk and sold it in Britain with the slogan 'Sir Hans Sloane's Milk Chocolate . . . for its lightness on the stomach and its great use in all consumptive cases'. The chocolate drink sold handsomely, though the snake fared less well. It was shot by the Duchess of Albemarle after it escaped the confines of a large pot. Sloane – who succeeded Sir Isaac Newton as President of the Royal Society and gave his name to a famous square in Chelsea – later accumulated numerous other collections of plants and curiosities from around the world, from opium and cannabis to Chinese rhubarb and insects of every kind. His treasure trove eventually became the basis for the British Museum collection.

* * *

Along with recreational drugs and minor stimulants like vanilla and chocolate, the tropical forests yielded many poisons that were new and alarming to Europeans. The first tentative explorations of the interior of Africa in the mid-nineteenth century yielded some that appeared almost magical and devilish to their collectors. The use of ordeal poisons by African witch-doctors particularly inflamed the passions. In 1846, a British Army doctor called William Daniell, who was travelling among the Efik people in West Africa, came across a creeper that produced a fruit known as the Calabar bean, containing the poison physostigmine. Daniell later described how it was used to establish the guilt or innocence of a person accused of some crime. 'The condemned person, after swallowing a certain potion of the liquid, is ordered to walk about until its effects become palpable. If, however, after the lapse of a definite period, the accused should be fortunate as to throw the poison from off his stomach, he is considered innocent and allowed to depart unmolested.' If he died, of course, he could be presumed guilty.

In Madagascar, the French came across many villages that conducted similar gruesome trials using the bean of the *Tanghinia venenifera* vine, the so-called 'ordeal tree'. The French were so appalled by the resultant suffering that they ordered the destruction of all Tanghinia trees on the island, though without great success. While their horror at the cruelty was well placed, their disbelief at its role in establishing guilt could be wide of the mark, for this approach to criminal justice is not as mad as it seems. Most ordeal poisons were chosen because they do not kill if the accused person takes them quickly enough and vomits them up – as might happen with someone confident of both their innocence and the powers of the potion as a truth drug. But anyone feeling guilty, who might sip slowly in the hope of a lucky escape, would probably not throw up and would surely die.

Very slowly, too, doctors and apothecaries in Europe began to consider the wisdom of ancient potions for curing diseases. Doyens of London's Royal Society in the seventeenth century began to believe that native botanical treatments for diseases like fever were sometimes preferable to conventional European treatments, which generally comprised bleeding, purging and the imposition of leeches. Tree barks from the forested foothills of the Andes were in use from the 1630s as a treatment for malaria, and soon after, Sloane popularized 'bamboo tarr', an ointment from the West African forest, to treat gout and rheumatism. Thus the lands of diseases and fever became also the land where the cures were to be found.

CURARE

Curare is one of the most potent and famous jungle poisons. The French scientist-adventurer Charles Marie de la Condamine first spotted it in use among Amazon Indians in the mid-eighteenth century. It was investigated further, half a century later, by the German naturalist Alexander von Humboldt and his French colleague Aimé-Jacques-Alexandre Bonpland. The latter pair were the first to realize that it is in fact a mixture of chemicals prepared from the deadly bark of several trees, with snake venom and poisonous ants sometimes added for extra potency. The mixture is then boiled for two days, strained and evaporated to produce a thick paste.

The best curare makers have always sold their potions to neighbouring tribes, who hunt with curare-tipped arrows or blowguns made of bamboo. Curare causes death by paralysing muscles until the animal stops breathing. (This was discovered in the 1820s by British adventurer Charles Waterton, when he poisoned an ass with curare then revived the 'corpse' by ventilating its lungs with a pair of bellows.) Frogs die in a few seconds, allowing curare preparers to test their potions by counting how many hops a poisoned frog can make before it drops dead. Birds die in a couple of minutes, but a big tapir might take twenty minutes to die. The great advantage of curare is that it poisons only the blood, so eating the meat is safe.

GIANTS OF THE JUNGLE

By the late eighteenth century, Alexander von Humboldt had dismissed the longings for an El Dorado to emerge from the undergrowth as part of 'the poetic imagination of mankind'. He took the simple step of going to the places where earlier adventurers had claimed to find the city of gold and the golden lake beside it. He found that the great lake was a mere cataract and the city did not exist at all. 'The illusion entertained for nearly two hundred years, which in the last Spanish expedition in 1775 cost several hundred lives, has finally terminated by enriching geography with some few results,' he concluded.

Humboldt's El Dorado was the wildlife. As he embarked on a five-year voyage to the Americas in 1799, he wrote of his vision of the tropical rainforest as 'an inexhaustible treasure trove' of nature. And he was not disappointed. His records report: 'animals of different nature succeed one another. Alligators appeared on banks, motionless, with their mouths open, while by them and near them capybaras, the large web-footed rodents that swim like dogs and feed on roots, appeared in bewildering herds, even lying among the alligators, seeming to know that these repulsive reptiles do not attack on land. Tapirs broke through the tall grass and slipped down to the river to drink.' It was, he reported to his native pilot, 'just like paradise'.

Later, he mused on how 'every object declares the grandeur of the power, the tenderness of nature, from the boa constrictor which can swallow a horse, down to the hummingbird balancing itself on a chalice of a flower', and how 'this aspect of animated nature, in which man is nothing, has something in it strange and sad'. During that ground-breaking trip to Peru, Ecuador, Colombia, Venezuela and Mexico, Humboldt and his French botanical companion Aimé-Jacques-Alexandre Bonpland collected some sixty thousand plant specimens, documenting the increasing evidence that the rainforests contain the most bewildering variety of plants and animals on Earth.

Bonpland, a surgeon who later spent ten years in a Paraguayan dungeon on charges of spying, may have done most of the collecting, but Humboldt had the grand vision. He produced one of the earliest and best manifestos for the emerging field of natural sciences when he wrote that 'the aims I strive for are an understanding of nature as a whole, proof of the working together of all the forces of nature'. And his eventual seven-volume, 3,700-page report on his journey inspired generations of future botanical explorers.

Bonpland and Humboldt produced the first accurate account of how to prepare the poison curare, which had first been spotted in use by Charles Marie de la Condamine half a century before. The two also discovered the electricity in the electric eel, and a rather tasty nut, later christened the Brazil nut. They set a world altitude record while climbing the Andean Chimborazo mountain, discovered the Humboldt ocean current off Peru and sailed up the Orinoco, where they discovered a geographical marvel unnoticed by previous explorers – anatural canal between the headwaters of the

Orinoco and the Amazon, where water would sometimes flow in one direction and sometimes in the other. They were, on the whole, less impressed with the people of the Amazon: Humboldt wrote how 'human nature is not seen here arrayed in that gentle simplicity of which poets in every language have drawn such enchanting pictures . . . these natives of the soil . . . their bodies covered with earth and grease, and their eyes stupidly fixed for whole hours on the drink they are preparing, far from being the original type of our species, are a degenerate race, the feeble remains of nations which after being long scattered in the forests, have been again immersed in barbarism' – a prescient remark, as we shall see.

For Humboldt, the jungle was not an impediment to his journey; it was his journey – and much the same could be said for Charles Darwin, Humboldt's greatest and most important disciple, who made his own round-the-world voyage aboard the *Beagle* just thirty years later. During his long, worldwide scientific expedition, Darwin spoke in strongly felt terms when he first saw the Brazilian rainforest. His journal one day in 1832 reports: 'Here I first saw a tropical forest in all its sublime grandeur . . . nothing but the reality can give any idea how wonderful, how magnificent the scene is . . . Delight itself is a weak term to express the dealings of a naturalist who, for the first time, has wandered by himself into a Brazilian rainforest.' It was, he wrote, 'one great, wild untidy, luxuriant hothouse'.

* * *

But this moment of pure delight and pure science was not to last. The British Empire, and the empires of the other European powers, were an increasingly important political and economic force, and the potions and botanical curiosities of the early years of exploration were destined to become increasingly valuable global commodities. The world was approaching its first era of economic globalization.

Thus Kew Gardens in London, which began in the eighteenth century as a simple royal pleasure garden, was transformed first into a scientific hothouse and then into the heart of a concerted, British imperial botanical endeavour. Plants gathered by gardeners and bureaucrats, chancers and traders, from every obscure part of the world were systematically collected and assessed for their commercial importance. Many of them were then distributed across the planet, through a growing network of imperial botanical gardens.

Kew's director for many years at the end of the eighteenth and the beginning of the nineteenth centuries was Joseph Banks. He was a trained botanist who had accompanied Captain Cook on his first voyage to the Pacific. Later, according to one senior official at the Colonial Office in Whitehall, he became 'the staunchest imperialist of the day'. He was a Privy Councillor, a trusted member of the inner circle of government and a founder member, in 1788, of the splendidly named Association for Promoting the Discovery of the Interior Parts of Africa. Early on, like many before him, Banks also showed a great interest in gold, believing that great hoards lay undiscovered in the jungle. 'Gold is found abundantly in all the torrents which fall into the Niger,' he wrote to parliamentarians in 1799 when proposing an imperial foray into West Africa. But as the years passed, he came to see the real treasure in the African forests as botanical and Kew as 'a great botanical exchange house for the empire'.

By the mid-nineteenth century, Kew had collected the best Manila and Havana tobacco, sent West African oil palm to India, Jamaica and Australia, sent the macadamia nut from Queensland to the West Indies, South Africa and Singapore, and distributed the pineapple and the eucalyptus tree across the globe. In 1844, William Purdie, one of Kew's gardeners, went to the Colombian jungle to find the ivory-nut palm, which wood-turners used for canes and umbrellas and as a substitute for ivory in billiard balls, buttons and chessmen. When a quarter of a million acres of coffee plants in Ceylon were devastated by leaf rust in 1873, Kew sent Liberian coffee culled from West Africa.

For the British, the greatest discovery among the forest plants came from the Far East in the form of tea. *Camellia sinensis* is native to the rainforests of northern Thailand, eastern Burma, northern Vietnam and the Yunan province of northern China. It was, of course, a precious commodity for the Chinese for thousands of years before British botanists discovered its mildly stimulating qualities. The British East India Company first shipped tea home in the early seventeenth century, but it was more than two hundred years later that tea plantations were set up in India and Ceylon to assuage the nation's growing thirst for the brew. By then its association with the rainforest was but a distant memory.

* * *

Darwin spent little time in the jungle: his greatest exploits were on small islands, such as those of the Galapagos, off Ecuador, whose small collections of unique species opened his eyes to the driving forces of evolution. But Victorian England produced three other giants of botanic jungle exploration: Alfred Russel Wallace, Henry Bates and Richard Spruce, who knew each other and corresponded regularly from their various jungle vantage points across the tropics. Occasionally these young men met, travelled together and parted crossly. All were corresponding, too, with Darwin, who was several years older and already a revered figure, living in an old parsonage in Kent – the tranquil garden county of England. There he nursed great ideas, but also a mysterious debilitating disease that had afflicted him since his own greatest journey.

Richard Spruce was a sickly, TB-suffering maths teacher who had an early obsession with the mosses and their close relatives, the liverworts, in his native Yorkshire. He arrived in South America in 1849 at the age of 32, intent on exploring the Amazon 'before it is too late' – meaning too late for him and his feeble physiology, rather than the forest. But he proved more durable than expected and spent the next eight years in the forest on one of the greatest botanical journeys of all time.

Spruce collected ferns, mosses, lichens and liverworts, some of which have never been seen since. Living off the land, making camp and carefully packing and labelling his specimens every night, he explored the swamplands of the Amazon's northern branch, the Rio Negro, many of which are still largely uncharted to this day.

THE SPICE ISLANDS

Kew was not alone in scouring the globe for potentially profitable plants – nor the first. The Portuguese and later the Dutch had long since staked out the East Indies, and especially the Moluccas islands between the Philippines and New Guinea. There, on the forested slopes of volcanoes, they found the clove and nutmeg trees that gave these islands their popular name of the 'spice islands'. Both clove and nutmeg preserved meat in the days before refrigeration and helped to disguise the bad taste if the meat did start to go off. The Dutch found that the nutmeg tree, which grows to 20 metres high, also provides mace from the casing that surrounds the nutmeg seed. And they learned from the Chinese how cloves could be a cure, or at any rate a disguise, for bad breath.

As early as 1512, while the Spaniards were chasing gold in the Americas, the Portuguese had seized some of the Moluccas islands and taken over the spice trade to Europe. It is said that when Ferdinand Magellan's ship reached home in 1522 after completing the first circumnavigation of the world, the entire cost of the three-year trip was recouped by selling the cloves it brought home. But it was the Dutch who turned the spice trade into a huge business, after capturing Portuguese forts on the islands a century later. At the height of the trade these spices were, literally, worth more than their weight in gold on the docks in Amsterdam.

By the late seventeenth century, Dutch botanists were busy collaborating with local plant collectors and doctors in Asia to identify useful plant medicines. Hendrik van

Reede of the Dutch East India Company produced a mammoth work called *Hortus Indicus Malabaricus* on southern India, and inspired Georgius Rumphius to do the same for the Indonesian island of Amboina. Back home in Amsterdam they sold plants for the preparation of remedies and they lectured doctors on them. And of course, exotic animals came too: the first elephant showed up in Amsterdam in 1679 at the request of William III of Orange, who was assembling a 'cabinet of curiosities'.

The Dutch traders built up the mystique of their cargoes, partly to maintain prices. But in the process they greatly increased the public's wonder at the miraculous nature of both the Far East and the jungle world of the tropics. Cinnamon, the Dutch traders claimed, would grow only in the centre of a mysterious jungle lake, guarded by great eagles and dragons. The Dutch monopoly on trade to the spice islands was finally broken by the French. In 1753, Pierre Poivre, the governor of the French-run Indian Ocean island of Mauritius, dropped by to see the plantations, secretly bagged some cuttings of cloves, nutmeg and cinnamon, and started his own plantations back home. The descendants of the pilfered clove cuttings eventually ended up on the island of Zanzibar, which became the largest producer of cloves on Earth.

Spruce then headed upstream into the forest regions of Peru and Ecuador, and altogether travelled ten thousand miles on the rivers of the Amazon basin, defying crocodiles, malaria and head-hunters. He survived constant fever, a near-drowning, starvation and two murder plots, during one of which he heard assassins devising ways to dispose of his body. Along the way, he collected thirty thousand botanical specimens, including seven thousand flowering plants new to science, and gathered the most complete collection of mosses in the world. All of these were eventually delivered – usually for a price – to European centres such as Kew. He needed to keep sending in bulk as he had no other income and received just two pounds per hundred specimens.

While never using his discoveries as a springboard for great theorizing about evolution, as Wallace and Darwin did, Spruce drew his own conclusions from his journeys. Rainforest species had 'a tendency to gigantism', he noted. 'Nearly every natural order of plants has here trees among its representatives. Here are grasses of forty, sixty or more feet in height [he had in mind bamboo and palms] . . . Instead of periwinkles we have here handsome trees exuding a milk which is sometimes salutiferous, sometimes a deadly poison.' There were 'violets the size of apple trees', bird-eating spiders, goliath beetles and giant snails, 8-inch slugs, 11-inch millipedes, butterflies with 7-inch wingspans and moths that can reach 12 inches across.

Spruce saw the value of local knowledge better than most and learned twenty-one different local dialects in his efforts to find out how the tribes used the plants – laying the basis for the modern science of ethno-botany. He collected foods from the forest and arrow poisons; he reported how the Indians made caffeine drinks from the seeds of vines mixed with cassava flour; and in southern Venezuela he reported that the Waika people mixed a paste made from the guarana berry with water and drank it 'first thing in the morning, on quitting their hammocks, and consider [it] a preservative against the malignant billious fevers which are the scourge of the region'.

Spruce was also on the trail of potent hallucinogenic snuffs such as yopo, which the Waika made by grinding the seeds of a leguminous plant called *Anadenanthera peregrine*. The powder made them twitch and tremble, apparently drunk, for around five minutes before they became 'drowsy while the devil, in their dreams, shows them all the vanities and corruptions he wishes them to see'. For taking the snuff, he recorded, 'they use an apparatus made of the leg bones of herons put together in the shape of the letter Y, the lower tube being inserted in the snuff box and the knobs in the nostrils'. Spruce also produced the first reports of perhaps the most violent hallucinogen of all, the ayahuasca or 'vine of the soul'. Soon after drinking a liquid made by boiling the vine bark, Indians saw 'beautiful lakes, woods laden with fruit, birds of brilliant plumage' and sometimes jaguars, and giant snakes that wrapped them in their coils, before paradise turned to panic as they 'turned deadly pale, trembling in every limb, bursting into perspiration and seeming possessed with reckless fury'.

Spruce sent back to Kew's then director, Sir William

Hooker, cultural artefacts such as devil dresses made of fibre from fig and Brazil-nut trees and worn at funerals; trumpets made from palm trees; and the snuff-taking apparatus of the Waika people. Like his contemporaries Humboldt, Darwin, Wallace and the rest, he saw the economic potential of scientific exploration in the tropics. 'How often have I regretted that England did not possess the Amazon valley instead of India,' Spruce once said. 'If that booby [King] James had persevered in supplying Raleigh with ships, money and men until he had formed a permanent establishment, I have no doubt that the whole continent would have been at this moment in the hands of the English race.'

But the reclusive Spruce was never part of the scientific establishment and never on first-name terms with civil servants and ministers. He had to make do in his retirement with a pension of one hundred pounds a year – and then only after a fight by Hooker to obtain it from Whitehall. Nor was he, at the last, a true botanical imperialist. He loved the rainforest and its inhabitants for themselves, and after travelling back and forth across the Amazon jungle, his greatest love remained the entirely uncommercial and uncharismatic liverworts. In a letter home from the Amazon, he wrote that 'it is true that [liverworts] have hardly as yet yielded any substance to man capable of stupefying him or of forcing his stomach to empty its contents, nor are they good for food; but if man cannot torture them to his uses or abuses, they are infinitely useful where God has placed them; and they are, at the least, useful to, and beautiful in, themselves – surely the primary motive for every individual existence.'

* * *

Spruce had much in common with his two fellow 'greats' of Victorian botanical jungle exploration: Wallace and Bates. All three were uncommonly shy, hiding in the jungle, perhaps, from both society and women; and they did not go in for the histrionics indulged in by many of the more bombastic and self-aggrandizing African jungle explorers who were to follow them. As Tim Severin in his *Spice Islands Voyage in Search of Wallace* put it, Wallace 'did not go forward, rifle in one hand, Bible in the other, at the head of a long line of porters, on the lookout for big game or souls to save'. While Livingstone and Henry Morton Stanley were taking huge expeditions across Africa, naming waterfalls and great lakes after their monarch, Wallace and Spruce usually travelled quietly and alone, with no firearm larger than a shotgun and winning the recognition and support of the natives (Wallace is famous and revered in Indonesia to this day, Severin reports). Wallace never played up the danger of his travels, which he claimed were no more dangerous than walking the streets of London.

Both, moreover, were without independent means and lived by selling their forest specimens. They were, by necessity, probably the first professional naturalists. Wallace was the son of an impoverished lawyer from the Welsh borders. At fourteen, he became a builder's apprentice and later a trainee land surveyor, drawing property boundaries as the common lands were privatized. By now a convinced socialist, Wallace described this creeping privatization of the land as 'legalised robbery of the poor'. In his early twenties, and out of a job, he took a post as an usher in a private school in Leicester, where he found time to read and chanced on books by Darwin and Humboldt. Here too, while browsing the natural history shelves in the city library, he by chance met a soul mate, Henry Bates.

Both men were ardent entomologists in their mid-twenties. Neither had backgrounds of privilege or education, but both yearned for the jungle. Within months, the two young men had concocted a plan to scrape up some cash and head for the Amazon. They left in 1848, planning to finance the trip by sending specimens back home. Wallace also vouchsafed that he would use his journey to provide data 'towards solving the problem of the origin of species', which was increasingly troubling the world of science. The pair docked at Belem and headed inland, sailing up to the town of Santarem, where they began work collecting plants. After two years, they tired of each other's company and parted – 'finding it more convenient to explore separate districts and collect independently', as Bates carefully put it.

Bates stayed at Santarem, and eventually spent eight years in the Amazon, diligently collecting samples of plants and insects, before he succumbed to a bad bout of malaria and headed home with fifteen thousand species, an astonishing eight thousand of them new to science. His journals are full of detailed descriptions of toucans and macaws, butterflies and giant spiders, piranhas and anacondas, sloths and army ants, manatees and marmosets. He documented cannibalism and a strange Indian custom of burying the dead, trussed up in jars beneath their huts. He described traditions of strange jungle creatures like a Bigfoot, equipped with backward-pointing feet to confuse trackers, a mythical river serpent called Mai d'Agua and dolphins that, mermaid-like, took the shape of beautiful women and lured men to their doom. But Bates was clear-eyed about the people and communities he met: he neither romanticized nor demonized. Take this from a visit to a remote settlement of Mura Indians in 1854: it was, he said, 'a miserable little settlement. The place consisted of about twenty slightly built mud hovels and had a most forlorn appearance, notwithstanding the luxuriant forest in its rear.'

During this time, Bates's extraordinary powers of observation brought some of the earliest insights into the complexities of rainforest ecology. He saw how extraordinarily the lives of individual species were connected, and how they must have evolved to fill their own niches in their lush world. His greatest discovery was how some insects have evolved to mimic one another. He noticed, through what must have been exquisitely painstaking observation, that one species of butterfly with highly distinctive colouring was in fact two species. One grew from a caterpillar that spent its days absorbing chemicals from a particular vine that made it toxic to birds; the other did not, and was perfectly edible. But the birds of the forest, unable to distinguish between them, left both alone for fear of eating the toxic one. Bates concluded that the second butterfly had evolved to find safety by mimicking the appearance of the toxic butterfly. He called the discovery 'a most beautiful proof of natural selection', and Darwin hailed it as a research triumph.

Wallace meanwhile, having left Bates behind to his butterflies, went west, up the greatest Amazon tributary of them all – the Rio Negro. There he recorded the now-famous 'cock of the rock', a bright orange pigeon-like bird with a pronounced crest. After four years in the Amazon, he returned to Santarem, where he met the newly arrived Spruce. The meeting amounted to a botanical equivalent of the famous encounter between Livingstone and Stanley; but it quickly turned tragic. Spruce had come out with Wallace's younger brother, Herbert, who promptly died of yellow fever. Wallace, himself racked with fever, returned to London but, on the way home, his ship caught fire and sank and all his specimens and notes were lost. Undeterred, he wrote up his notes from memory and, in 1854, started afresh, this time heading for Malaya and all points east. There he stayed for eight years, visiting virtually every inhabited island in the East Indies.

* * *

'I was in a new world,' Wallace wrote on his arrival in the east. 'Few European feet had ever trodden the shores I gazed upon; its plants and animals and men were alike almost unknown, and I could not help speculating on what my wanderings there might bring to light.' They brought to light a lot: while various Portuguese and Dutch feet had trodden these beaches over the years, their interest had been largely confined to a few commercial spice crops. Wallace had a much broader canvas and was soon shipping home bird skins, butterflies, beetles and much else from the East Indies.

(He sent regular shipments this time, to avoid a repeat of the Amazon disaster.) He was the first Briton to see birds of paradise in the wild, the first white man to see the elaborate courtship displays of the males, and the first to bring live birds of paradise back to Europe. He collected 127,000 specimens and made a series of fundamental discoveries about biology and geography that formed the basis for his book *The Malay Archipelago*, one of the greatest ever works of natural history.

Wallace also set to thinking here about the conundrum behind the origin of species. Many scientists of the day were aware that the species of the planet probably changed and may well have evolved one from another, but nobody could work out quite how, and many baulked at the suggestion that humans, too, might be part of this process. Darwin had been brooding on the idea for years; the slow pace of his progress was partly due to his fears about the explosive impact of his revolutionary ideas, and partly to a painful tropical disease from which he had suffered since his time on the *Beagle*, and which left him frequently lethargic. (It was probably Chagas' Disease, a form of sleeping sickness picked up from insect-borne parasites in the Brazilian forest, though it had affected Darwin so much that many thought his condition psychosomatic.)

Wallace knew that Darwin had been worrying away at the problem for years, but without much published result. The two men corresponded from time to time, though more about Wallace's finds than any grand theories. But in early 1858, somewhere in the Moluccas islands, Wallace had what amounted to a Eureka moment during a bout of malarial fever. When the fever subsided, he wrote up his insight in a four-thousand-word article called 'On the Tendency of Varieties to Depart Indefinitely from the Original Type', and sent it off by packet ship to his friend. Darwin was astounded both by the fact that Wallace had reached similar conclusions on natural selection to those that he was painfully developing, and at the vigour and clarity with which the younger man had set them down.

Wallace's ideas about evolution were in fact subtly different from those of Darwin. While Darwin argued that it was the cut-throat competition between individuals that ensured the 'survival of the fittest', Wallace saw fitness more in environmental terms: species survived or failed according to their ability to fit into their environment, he said. It could have been war between them, but instead the mutual friends agreed to compile a joint paper, which was presented on their behalf at the Linnean Society in London in 1858. Neither attended: Darwin was laid up with his usual trouble while Wallace was still in the east.

While Darwin was ever afterwards punctilious in acknowledging Wallace's role in the theory, he was nonetheless galvanized into action by Wallace's paper and worked night and day to get his full version – a very long work entitled *The Origin of Species* – published before Wallace could produce any rival. Darwin's work was already a best-seller before Wallace stepped on to dry land in Portsmouth in 1862 with two birds of paradise and a final collection of specimens. With the joint presentation to the Linnean Society soon forgotten by all but a few scientists, Darwin's best-seller ensured that he was seen as the author of the idea. He swiftly became one of the most famous and controversial people of his day, lauded but also reviled as 'the devil's chaplain' and worse.

Wallace never complained at being eclipsed, telling Darwin that 'I shall always maintain that the idea was yours'. He probably never wanted the limelight and by now was as much interested in spiritualism as he was in evolution. Darwin was laid to rest in 1882 in Westminster Abbey after what amounted to a state funeral, but before he died he wangled the penurious Wallace a small pension. Wallace lived to enjoy it, pottering in his Dorset garden for another thirty years.

THE WALLACE LINE

Wallace's journeys among the islands of the East Indies brought one final great discovery that opened up a debate that continues to this day on the high diversity of species in the rainforests. He noticed a hitherto hidden biological divide in the archipelago in which, as he put it, 'one half shall truly belong to Asia, and the other shall no less certainly be allied to Australia'. The divide was very precise, following a line between the islands of Bali and Lombok. This divide is geographically minute – the two islands are separated by a channel just 25 kilometres across – but biologically it was and remains immense. In Bali, Wallace found Asian birds such as weaverbirds and wagtail thrushes and white-plumed starlings. None turned up in Lombok, where the forests were full of honeysuckers and pigeons and turkeys. Equally he saw a divide among the animals. In Borneo, on the Asian side, there were monkeys and large cats and wolves and bears and hyenas and elephants. But in Celebes, New Guinea and all points east there were none of these. Instead, they had the marsupial kangaroos and wombats and the platypus that are, unmistakably, Australian.

Between them, invisibly, was what we now know to be a fault line in the great continental plates. In times past, these two zones were far apart. Wallace did not know this, but he saw clearly that life evolved independently, for the most part, on each side of the divide. Today we know that divide as the Wallace Line.

The Wallace Line is not the only hidden fracture in the world's web of life. Just as remarkable is that between the

African mainland and the island of Madagascar. The latter, as we now know, broke away to form an island some 70 million years ago and subsequently evolved its own rainforest ecology. As a result, it has a higher proportion of species unique to the island than any remotely comparable area on Earth. When French naturalist Philibert Commerson visited the island in 1771, he called it 'the naturalists' promised land. Nature seems to have retreated there into a private sanctuary, where she could work on different models from any she has used elsewhere. You meet bizarre and marvellous forms at every step.'

Commerson found a 'spiny forest' occupied by dense thickets of green spikes armed with giant thorns, quite like cactuses in appearance but genetically entirely different. There was a huge range of lemurs, a whole class of animals found nowhere else but here, ranging from mouse-sized animals to evidence of gorilla-sized lemurs that existed in prehistoric times. Many species survive on the island mainly because there are no large predators like lions or tigers. Madagascar is the home, too, of the chameleon and the rosy periwinkle. This small flowering plant, traditionally used all round the Indian Ocean as an infusion to ward off diabetes, became the subject of a pharmacological gold rush in the 1960s after scientists discovered that it contained alkaloids that arrest both Hodgkin's Disease and childhood leukaemia.

DARWIN'S LAST LAUGH

Even at his death, Darwin's ideas of natural selection remained controversial. For many sceptics, the confirmation of his theories came only with the strange story of the comet orchid, a native of Madagascar. The orchid produces white waxy flowers that attract pollinating insects with a fragrance that is specially seductive after dark; but any insect alighting on the flower in the forest night will have great trouble finding its nectar, because it is hidden deep inside a narrow tube roughly 30 centimetres long. This curious plant was first discovered by Aubert du Petit-Thouars, a French aristocrat and botanist who narrowly escaped the guillotine during the French Revolution and spent years travelling Africa before daring to return home. It was another fifty years before the first comet orchid was coaxed into flowering in Europe, in 1857, and a further five years before its notoriety reached the attention of Darwin at his home-cum-laboratory in Kent.

For Petit-Thouars, the plant was little more than a curiosity. For Darwin, it raised a fundamental question: what kind of insect could suck the nectar from such a flower? He experimented in his greenhouse, using bristles and needles and glass rods, but the answer was clear enough to him. For this orchid to be pollinated, it required an insect with an extremely long proboscis, many times the length of its body. Darwin predicted that one day such an insect would be found. He was laughed at by many of his peers for what seemed an outrageous suggestion. Whoever heard of a moth with a tongue as long as that of an elephant? And yet, another forty years later, in the depths of the Madagascan jungle, a subspecies of the Morgan's sphinx moth was discovered with just such a tongue – and it pollinated the comet orchid. Darwin was long dead by then, but he surely had the last laugh.

HEARTS OF DARKNESS

From peanuts to rubber and from anti-malarial drugs to the spices of the East Indies, jungle plants have given us some of our most important substances. It is impossible to imagine the nineteenth-century imperial conquests without the bark of an obscure shrub from the foothills of the Andes; or the transport revolution of the twentieth century without the strange milky excrescence from a tree found deep in the Amazon. But our exploitation of this cornucopia often instigated a reign of terror in the jungle. It was the white man who brought a 'heart of darkness' to Africa – and yet we could still go in search of the forest's simple delights, like the tastes of cinnamon and nutmeg and vanilla from the Moluccas islands and Mexico, and the exotic plumes of birds of paradise from New Guinea. Even today, scientists are uncovering new, exquisite orchids from the South American cloud forests.

JESUIT'S BARK

Sometime early in the seventeenth century, as Europeans tightened their grip on Latin America, Jesuit missionaries in the Peruvian Amazon stumbled on a native forest treatment for fever. The plant became one of the most valuable ever discovered, and, to this day, forms the basis for the treatment of the deadly tropical disease of malaria. It is no exaggeration to say that, without it, much of the story of imperial invasion of the tropics and its rainforests would be different.

The tree that changed the course of history was named after Lady Ana, the Countess of Cinchona and wife of the viceroy of Lima. In 1638, according to legend, Lady Ana became the first European to be cured of malaria using an infusion made from the tree's bark. The bark had been supplied by Indians living 800 kilometres away from Lima in the rainforests of Loxa, in what is modern Ecuador. When the recovered Countess went home to Spain, she took the bark with her and – so the story goes – began using it to cure fevers.

The story of the Countess's cure, though still widely told, seems to be a myth, got up by an Italian doctor a generation after Lady Ana's death and embellished later by Victorian scientists in Britain as part of their own myth-making. Detailed Spanish court records dug up by historians give no such account, and the Countess herself appears to have died a decade before the events described. But, even stripped of the romantic tale, it does seem that the Jesuits of Peru somehow learned about the bark and its magical powers. Native Americans may not have suffered from malaria till the arrival of Europeans, but they certainly suffered from similar vector-born fevers and fought them with the bark of the tree that became known as cinchona.

The earliest authenticated European report of the bark's medicinal qualities comes from an Augustinian monk called Antonio de Calancha. He wrote in the 1530s that 'in the country of Loxa grows a tree which they call the fever tree, whose bark, of the colour of cinnamon, made into powder and given as a beverage, cures the fevers'. According to Jesuit records, the bark was first brought back to Europe in 1643 by Father Bartolome Tafur, and was thereafter promoted by Cardinal John de Lugo. Its use spread, particularly in southern Europe where there were regular outbreaks of lethal fevers – most notoriously in Rome, where several cardinals were afflicted whenever they assembled in the marsh-surrounded city to choose a new pope.

But, in the fevered political atmosphere of northern Europe at the time, the new medicine's popularity was not helped by its association with Jesuits. Protestants believed for a long time that the spreading use of the bark was a papist plot and, because the right dose of bark was uncertain, it did sometimes kill through an overdose, and at other times failed to cure at all. Some called the bark the 'Jesuit's poison' – and such fears may have persuaded the fanatically Protestant Lord Protector of England, Oliver Cromwell, not to take it before he died of fever during an epidemic of malaria in Britain in 1658. Had he taken the bark and recovered, British history might have been very different.

An early enthusiast for the treatment was an obscure apothecary from Cambridge in England called Robert Talbor. He prescribed potions based on cinchona bark to the fever-struck people of the Essex marshes and, in the process, perfected a secret remedy that he brought to London and began selling to the rich and famous. The medical establishment, who preferred leeches and purging, derided him as a quack, but he cured Cromwell's successor as head of state, King Charles II and then, with royal patronage, followed up by successfully treating the son of Louis XIV in France and sundry other European aristocrats during a major epidemic in the late seventeenth century.

Talbor went to great pains to keep secret the recipe for his concoction, even warning against the use of other 'palliative cures, especially Jesuit's powder'. It was only after his death in 1681 that it emerged that his own cure was in fact based on the dreaded powder, which he appeared to have bought secretly for many years from a seminary in Belgium. The religious stigma was lost and the bark from the Andean jungles soon became the potion of choice against malaria throughout Europe. It was the most complete triumph yet for a jungle potion and did much to establish in European minds the potential value of a forest pharmacy.

From then on, demand for the bark soared, and ensuring sufficient supply became a major problem. The Indians and their Jesuit collaborators were able to maintain a monopoly on trade in the bark by keeping secret all details about the tree from which it was cut, and about precisely where and how that tree might be found. The secret persisted for 150 years. Before 1780, the only bark exported from South America was harvested in Loxa and shipped from the Peruvian port of Payta to Panama, and thence to Europe. The Jesuits are said to have required that native bark hunters plant a new tree for every one destroyed, but this did not always happen; the German naturalist Humboldt, who passed through Loxa in 1795, reckoned that twenty-five thousand trees of the species *Cinchona officinalis* were being killed annually to supply the European market, and warned that the harvest threatened future supply.

* * *

Meanwhile, the mystery of how the bark worked increasingly intrigued scientists. In the early nineteenth century, biochemists began to work on finding the vital ingredient in the bark. Finally, two young Parisian biochemists, Joseph Pelletier and Joseph Caventou, who had already discovered chlorophyll and strychnine, identified an alkaloid that they named quinine after the native name for the bark – quina. The alkaloid, it emerged, interferes with the growth and reproduction of the parasite that carries malaria, so halting the disease. Until then, the trade had been riddled with quackery and some bark sold in Europe contained little or no quinine at all. But until the active ingredient was discovered, there was no method of knowing which would cure and which would not. It emerged that a range of Andean shrubs of the cinchona family contained quinine, and, even within species, the quinine content varied dramatically with location. It then became possible for the first time to test bark and eliminate the element of chance from harvesting, purchase and prescription.

It later became possible to extract the alkaloid from the bark and give precise doses; and thus, too, arose the perfection of tonic-water – the mixer drink that, so the story goes, became popular after British soldiers in India discovered that the excruciatingly bitter quinine they took to counter malaria could be made more palatable by mixing it with gin. Eventually, they grew so fond of the taste of the two combined that manufacturers added much smaller doses of quinine to carbonated water to be drunk with the gin.

But, however the quinine was taken, it still required good bark as the basic ingredient. By the mid-nineteenth century, as European soldiers, traders and administrators spread through the tropics in ever greater numbers, demand for the bark was soaring. In 1840, annual sales of cinchona bark from South America to Europe passed a million pounds in weight. By now it was clear that trees outside Loxa also contained quinine; the region lost its monopoly and a lively trade grew up in Bolivia. But even so, supply could not meet demand, and the British, Dutch and other colonial powers were losing ever more empire-builders to malaria. They desperately wanted a secure, cheap and plentiful supply of 'fever bark' that cut out the middlemen. Victorian plant-hunters began to raid the rainforests looking for samples to bring home and propagate on tropical plantations close to their predominantly Asia-based customers. Colonial horticulturalist John Forbes Royle wrote that 'after the Chinese teas, no more important plants could be introduced into India'.

Put simply, they wanted to steal some cinchona seeds. Would this be dishonourable? Some said so. But the new generation of colonial botanists at Kew and elsewhere had an argument we are more familiar with today: the environment. The American natives and Spanish traders, they said, were driving cinchona to extinction and, if this carried on, the species – so vital to the future of the colonies – were at risk of becoming extinct. The botanical resources of the tropics were too precious to risk this, they said, and it was their duty to pluck them from the Amazon forests. 'The right of these races to remain in possession will be recognized; but it will be no part of that recognition that they shall be allowed to prevent the utilization of the immense natural resources which they have in charge,' said Benjamin Kidd, an Anglo-Irish civil servant and amateur botanist in his influential book of the time, *The Control of the Tropics.*

In the annals of Victorian botany, the credit for the eventual botanical larceny of cinchona bark is usually given to Clements Markham. An ambitious and well-travelled official with Britain's India Office, he offered to go to Peru, an old stomping ground, to collect cinchona seeds for the Empire. And he recruited several famous British Victorian botanists, including Richard Spruce, to the enterprise, which got off the ground in the late 1850s.

But the problem was finding the right plants in a very large jungle. There were known to be more than twenty species of cinchona, each looking rather similar, with flowers like lilacs. They grew in the botanical profusion of the cloud forests on the edge of the Amazon basin, where almost every hillside had its own indigenous species. But, while many species contained some quinine, few, if any, outsiders knew which were the best – and the handful of locals in the know were not telling.

Nonetheless, after months of rooting around in the Andean foothills from Bolivia to Colombia, Markham and his fellow conspirators brought out a selection of seeds that were later shipped to India and planted out in the hope of establishing viable cinchona plantations. There was much imperial crowing about this patriotic and courageous undertaking, but in truth nobody knew if the seeds were any good. They were certainly not the first. Several European botanical gardens, including Kew, had already received cinchona seeds in 1849 from a French collector called Algernon Weddell, and it is far from clear that Markham and Spruce found any better. While their plants did provide quinine, and eventually formed the basis for plantations in India that kept British troops and others supplied for many years, their yield of quinine was poor and the plantations always ran at a loss.

None of this deterred Markham, who revelled in his

Boy's Own triumph, playing up his daring role in collecting and smuggling the seeds, and adding a little spice by popularizing and embellishing the hitherto obscure story of the Countess of Cinchona. It was good PR. He was later knighted and became president of the Royal Geographical Society, where he was responsible for promoting another great British adventurer, Captain Robert Scott, who failed to become the first man to reach the South Pole, but died a hero in the attempt.

* * *

Markham's over-confidence about his seeds led Kew to one of its great commercial disasters. Complacent in the belief that it already had cinchona seeds growing in its colonial plantations, the garden's curators took no notice when another British plant collector offered them some more. Charles Ledger lived a colourful life on the shores of Lake Titicaca in Peru, where he was an alpaca-wool trader and sometime smuggler of plants and animals in the Andes. Once he infiltrated a flock of more than a thousand llamas over the desert border into Chile, from where they were exported to Australia. Noting the British search for cinchona trees, he spotted a chance to make a buck. He asked one of his most loyal helpers, an Indian plant collector named Manuel Incra Mamani, to find and bring him some cinchona seeds.

Mamani was an expert on cinchona trees and, Ledger later claimed, could distinguish twenty-nine different types. The two men were inseparable, but while Mamani was happy to collect bark for sale, he repeatedly refused to find cinchona trees for Ledger. They quarrelled about it and one day, as Ledger later told it, Mamani abruptly left his service and disappeared for four years. For at least some of that time, Mamani travelled into a stretch of Amazon rainforest in northern Bolivia, near the torrential headwaters of the Rio Beni. This was far from the forests where the Spanish collected their bark but, when he returned, he carried with him 6 kilograms of tiny seeds hidden in a pocket in his braids. They were, he told Ledger, from the best family of the best species of cinchona, called *Cinchona calisaya.* More than that, he claimed they all came from a patch of fifty trees that were the best of the best family. Their superiority was signified by a slight colouring of the leaves. 'No botanist in the world would ever have found them,' he said – and who could have disagreed?

Knowing his man, Ledger sent the seeds to his brother George in London for urgent dispatch to Kew. But the curators there turned down his offer of the seeds, declaring that Markham and Spruce had produced all the seeds they needed. So, in December 1865, George Ledger sold them instead to the Dutch Consul-General in London, for six hundred gulden. It was one of the best trades in history.

The Dutch, more in hope than expectation, shipped the seeds to their botanical gardens at Bogor in Java. There, horticulturalists planted them and grafted the seedlings with another faster-growing cinchona species to form the breeding stock for a quinine plantation. The new plant was given the name *Cinchona ledgeriana*, after its provider. When the plants matured, the bark turned out to be many times better than the seeds collected by Markham and Spruce – and, indeed, far superior to the bark harvested by Indians from Loxa region and supplied to the Spanish for two hundred years. They were, as Mamani had promised, the best, containing more than ten times more quinine than any other known plants. It was a botanical discovery of major significance for tropical health. For the first time, local Asians were supplied with quinine from these super-yielding plants, and Mamani's pocketful of seeds probably saved millions of lives over the succeeding decades.

By 1880, the Andean forests were churning out almost 10,000 tonnes of bark a year, but supply was becoming increasingly difficult. This was the moment when the plantations of the Far East, and particularly the Dutch operation, took over the market. Soon, nobody wanted to bring bark from the Americas. The British plantations in India and Ceylon were kept going and supplied local imperial staffs and favoured Indian workers, but the enterprise was strictly unprofitable, and it was the Dutch who dominated world trade for the next seventy years, until artificial versions of quinine were synthesized after the Second World War.

The scandal of Kew's refusal of Ledger's seeds was successfully buried. Ledger remains a much less well-known figure today than Markham: he has been entirely left out of several histories of the cinchona saga, merits not a line in the *Dictionary of National Biography* and was never rewarded for his efforts.

And Mamani? His journey into the Bolivian forests could reasonably claim to have yielded the most important botanic discovery of all time. He was the purveyor of the seeds that finally provided Europeans with profuse quantities of the chemical they needed to travel the tropics safe from the scourge of malaria. But in his homeland he was a traitor, an accomplice in smuggling a vital resource from his country – an agent of European botanical imperialism. Shortly after Ledger sent his seeds to London, Mamani's crime was uncovered and he was packed off to gaol in Coroico, Bolivia. There he was beaten, half-starved and tortured in an effort to get him to say to whom he had given the seeds. He refused to divulge his secret, and was eventually set loose to die. We still do not know exactly where he gathered his seeds, but, once they were successfully propagated in Java, few cared.

RUBBER GOODS

If cinchona bark is the greatest medicinal discovery from the jungles of South America, rubber is its greatest industrial product. First news of a strange, milky latex that natives obtained from trees by cutting incisions in the bark reached Europe after Columbus's second voyage in 1493. Columbus called it 'caoutchouc', from an Indian word meaning 'the wood that weeps'. The Indians used the latex as a crude method of waterproofing cloth – an invaluable attribute in a rainforest – and in Mexico, Hernando Cortez noticed the

SEARCH FOR THE ULTIMATE PEANUT

Whenever you grab a handful of peanuts, ponder this. The future of the world's most ubiquitous nut is under threat because modern botanical prospectors cannot find the whereabouts of what has become known as 'the ultimate peanut', the nut from which all today's peanuts grew.

The peanut's origins lie in the Amazon rainforest, in eastern Bolivia, just a few hundred kilometres from the heartland of Mamani's 'best of the best' cinchona bark. It has probably been cultivated there for upwards of four thousand years. And over the millennia, farmers and later scientists have bred peanuts to taste good and yield bumper crops in peanut (or groundnut) plantations. But since leaving the jungle, the peanut has lost most of its natural genetic ability to fight pests and disease. Peanuts are at a constant risk of some new epidemic, which will probably come out of the peanut's homeland in the Amazon jungle.

The only solution, says David Williams of the International Plant Genetic Resources Institute in Cali, Colombia, is to go back to the jungle and track down the 'ultimate peanut', the wild peanut plant, for that alone will contain the lost genes for resistance against the peanut's natural enemies. Williams wants to find those genes and add them to the genetic makeup of our modern cultivated varieties – either through conventional breeding or by GM technology. 'We think the ultimate peanut is out there, and it is sure to be endangered. I just hope we can find it in time,' he says.

But there is a problem: he has narrowed down the

search area to the Gran Chaco region of southeast Bolivia because that, he believes, is where the ultimate peanut must live. But he cannot get there. The Gran Chaco forest is virtually under siege at present. Oil companies have been building a pipeline through the forest against the vehement opposition of local fishing communities and, in this battle, foreign scientists are regarded by the fishing communities as one of the enemy. Fearful of exacerbating the conflict, the Bolivian government won't let the scientists in. But the peanuts won't wait.

Aztecs playing a game with a rubber ball. But the stuff remained, at most, a botanical curiosity and it was another two centuries before its true potential was spotted by Charles Marie de la Condamine, a Jesuit French geographer. La Condamine had been sent to Peru by the French Académie des Sciences in the 1730s to contribute to a project to measure the curvature of the Earth's surface and gauge its true size, but he filled in his time on botanical explorations in the Amazon, and has some claim to have written the first scientific account of an Amazon journey. He provided the first description of the South American poison known as curare, and also sent back samples and profuse details about how the Indians extracted and used the strange latex. 'They make bottles of it in the shape of a pear, to the neck of which they attach a fluted piece of wood. By pressing the bottles, the liquid is made to flow out through the flutes and, by this means, they become real syringes.' From this, the Portuguese called the tree 'pao de Xiringa' (syringe wood) and the rubber tappers gained the name they still have today, '*seringueiros*'.

La Condamine showed his find to a British friend, Joseph Priestley, who was both a radical theologian and a chemist with a happy knack of discovering things people wanted. He was already famous for having discovered oxygen, and he fathered the modern soft-drink industry when he spotted that carbon dioxide bubbled into water makes a pleasant carbonated drink. He played around in his lab with the new latex substance and noticed that, when hard, it would rub out pencil marks. He called it, with admirable directness, a rubber, and the name stuck.

The final decades of the eighteenth century and the first of the nineteenth century were a ferment of industrial innovation. There were rubber boots and catheters; the French made the first hydrogen balloons with rubber-proofed silk and boasted the first rubber factory to make elastic bands. Interest in the new forest product grew after a series of industrial discoveries in Europe and North America transformed rubber and its potential uses. Its pioneers became household names and nouns in the dictionary.

Thus Charles Macintosh, an industrial chemist employed in the Lancashire iron and steel business, gained immortality when, in 1823, he patented a system for mixing rubber with naphtha, a waste gas given off when distilling tar. This unlikely combination of sap from a tree in the Amazon jungle and a waste industrial gas from England dramatically improved rubber's waterproofing qualities. He proposed that by sandwiching a thin smear of the substance between two layers of fabric you could make light cloth fully waterproof. Within months, Macintosh was contracted to supply waterproof kit-bags, air-beds and pillows for one of the ill-fated Sir John Franklin's journeys to the Arctic. And, in less than a year, manufacture of the rubberized Macintosh raincoat was underway at a mill in rain-soaked Manchester – then the hub of the British industrial revolution.

But rubber was still a difficult product: despite its many qualities, it tended to become stiff in the cold and sticky in the heat. During the 1830s, however, a bankrupt American inventor called Charles Goodyear began experimenting with rubber. He was at the time incarcerated in a debtor's prison but still contrived to have access to basic laboratory equipment. One day, while mixing rubber with sulphur, he got his mixture too close to a hot stove and noticed that afterwards the mixture remained flexible when cold and dry when warm. Hey presto! He had invented a process of treating rubber that dramatically improved its value for a host of uses, and called the process vulcanization, after the Roman god of fire. The discovery didn't do Goodyear's finances much good: his patents were constantly infringed and his efforts at salesmanship ran up further debts. But half a century after his death, his name was given to a new brand, Goodyear tyres.

* * *

By the 1840s, Amazon rubber latex was being sent raw to London and other industrial centres to make everything from fire hoses and inflatable beds to catheters and contraceptives, which quickly garnered the euphemism 'rubber goods'. At the Great Exhibition in London in 1851, there was a large display of solid rubber tyres. And, as demand for rubber products grew, European and US industrialists rapidly squeezed out their Brazilian rivals, and Belem shoe manufacturers and waterproofers shut down. The Europeans also sought greater control over supply of the magical new material: no longer content to rely on haphazard deliveries from Brazilian rubber traders, they set up trading stations on the Amazon and its tributaries and sent poor migrant labourers from the drought-prone northeast of Brazil into the jungle to harvest ever more latex. They bought the products of the local tappers at ridiculously low prices, sustained often by systems of debt bondage not so different from slavery. But this was commerce: they were industrializing the rainforest. The new hell of Europe's satanic mills was being transported to the satanic forests.

In the US, the Amazon was seen as the 'new South'. They talked of sending black labourers from their own plantations to provide extra labour for the rubber business. The American Navy eyed the Amazon basin and the navigability of its river, sending survey ships upstream in the 1840s. For a while it was the destination of choice for adventurers. In 1855, before he ever made his name as the chronicler of Tom Sawyer and Huck Finn, Mark Twain stood on the docks of New Orleans asking about ships headed for Belem. 'I was fired with a longing to ascend the Amazon and to open up a trade with all the world,' he wrote in an essay entitled 'The Turning Point in My Life'. No boat came, however, and instead he hitched a ride with a Mississippi river pilot and learned about that river instead.

In the 1850s, as the rubber boom intensified, agents went further and further west, into the headwaters of the Amazon. They found that forests in the state of Acre and over the border in Bolivia contained the biggest stands of the three main Amazon rubber-bearing trees – *Castilla elastica, Hevea brasiliensis* and *Manihot glaziovii*. Acre became, for a while, one of the most valuable slices of real estate in the world. But the route down the Amazon to the Atlantic coast was long and tortuous. Thousands of lives were lost and fortunes wiped out trying to construct a railroad over the Andes to the Pacific coast. In the 1880s, with tens of thousands of migrant rubber tappers in the forests of Acre, a syndicate of bankers and US rubber barons sought to annex the state from Brazil and turn it into what amounted to a US colony under the nominal authority of Bolivia. It failed.

Large despotic colonies formed in these wild outposts. Perhaps the worst was run by Julio Cesar Arana on the Putumayo river on the borders between Colombia, Peru and Ecuador. Here, in an area the size of Belgium, the Peruvian trader employed Indians in chain gangs and strung them up or burned them alive if they failed to produce their quotas of latex. There was systematic rape, murder and torture. Thousands starved as Arana's men destroyed crops to force the Indians into virtual slavery. The women were kept in breeding farms. By the time the camp was finally shut down in 1914, amid a growing international scandal, an estimated fifty thousand Indians had died in the terror.

The height of the Brazilian rubber boom came in the final years of the nineteenth century after a vet from Northern Ireland called John Dunlop sought a way to prevent the jarring caused by the hard rubber tyres on his young son's tricycle when he rode it over the cobbled streets of Belfast. In 1888, he invented the pneumatic tyre – or, rather, reinvented it, for it turned out that the idea was first thought of forty years previously, but had then been entirely forgotten. But, in the business of ideas, timing is everything. His idea both set off a bicycling boom and enabled the motor car to become something other than the bone-shaker of all time. Dunlop, like Goodyear before him, made little money from his invention, however, and went back to his veterinary practice, which mainly involved keeping horses on the road. But others made their fortunes and his name lives on in the product he invented.

As the motor car began to make its mark, millions of pneumatic rubber tyres required tens of thousands of tonnes of rubber. Brazil's exports of latex in 1900 totalled 27,000 tonnes, a thousand times more than in 1831. In those mad days, Manaus, the metropolis at the heart of the Amazon, was known as the Paris of the Tropics. It boasted an opera house, the Teatros Amazonas, which was copied from La Scala in Milan, with a Byzantine golden dome above a Renaissance façade and four tiers of boxes with golden balconies. At its height, Manaus supported eight daily newspapers and – so the story goes – the town's high society sent their laundry to Lisbon and Paris for cleaning. It was no exaggeration to say that the industrial economies of Europe and North America ran on Brazilian rubber: without it, they would have ground to a halt. Despite the ostentation in Manaus, however, most Brazilians saw little benefit from this industrialization of the forests. But their suffering was nothing compared to what was happening in Africa . . .

PARADISE DISCOVERED

When the *Vittoria* sailing ship limped into the Spanish port of Seville in September 1522, those on the dockside recoiled in horror. Its sails were in tatters, the mast splintered and the crew half-starved. It was all that was left of a fleet of five ships that had set sail three years earlier under Ferdinand Magellan, the man who named the Pacific Ocean on the first circumnavigation of the globe. Of 277 men who set out, only nineteen returned. Magellan was not among them. He had been killed by natives in the Philippines even before the intended destination of the spice islands had been reached.

The returning ship brought rich treasures from the East. Among them, besides spices from the gardens and forests of present-day Indonesia, were some exquisitely coloured bird skins with long delicate plumes that had been given to the ship's captain by the Sultan of the island of Batjan, who had received them from New Guinea. Though crudely preserved and dried, nothing like them had been seen in Europe before. Their feet and all flesh had been removed, and the Europeans were spooked by this. They reasoned that, being without flesh, they must be ethereal creatures, and being without feet, the birds could not have landed on ground or tree. They called them 'birds of paradise'.

The birds were rare and revered in their own lands, where the plumes were often used as currency. But they were still rarer in Europe, where occasional examples arrived aboard ships returning from the East. No European ever saw a bird of paradise in the wild until 1824, when the French naturalist Rene Lesson described his first sighting as 'like seeing a meteor, cutting through the air, leaving a long trail of light'. It eventually emerged that the birds, with their outrageous plumages, were almost all from one island – New Guinea, the biggest, most impenetrable and most forested of all the islands of the Far East. They live high up in the forest canopy, where the males devote several months a year to conspicuous and acrobatic displays of their plumage, usually at dawn. The best are rewarded with queues of mates; the rest have no chance, and natural selection does the rest to ensure their finery.

Another nineteenth-century fan of the birds, whose nearest relatives are, oddly enough, the scruffy crows, was Alfred Russel Wallace. He was the first European to see their courtship dances in the canopy, and when he caught a boat back from the Far East in 1862, he brought two birds of paradise with him, feeding them cockroaches all the way. Once ashore in England, Wallace handed the birds to London Zoo, where one lived for two years – a rare example of a bird of paradise successfully surviving in captivity away from its native land.

The height of interest in these spectacular birds came in the nineteenth century when thousands were slaughtered to provide plumes for decorating the hats of Victorian ladies. The trade attracted young adventurers – including, at one point, the young Errol Flynn. At the same time that the birds' plumes became an essential fashion accessory, a bizarre trade grew up in Europe among some of the continent's richest men to obtain the most spectacular remains of the rarest birds, many of them unexpected one-off hybrids. The majority of these collectors' items ended up in the hands of the eccentric Lionel Walter Rothschild, of the Rothschild banking family, who had created a huge natural history museum in his home at Tring in Hertfordshire, north of London.

The plumage trade finally died in the early twentieth century, banned in the US and, by similar legislation, soon after in Europe. To this day, the birds are rarely seen in zoos, because they do not survive long in captivity, despite some dramatic efforts. Captain Ned Blood, a border patroller in the British sector of New Guinea in the 1940s, collected birds of paradise while upcountry among the highland tribal communities. He set up an aviary station at Nondugl, in the highland interior, from where he sold a few specimens to European zoos and aviaries.

But the best effort was made by another eccentric British collector, Sir William Ingram. In 1898, he bought the island of Little Tobago, a tiny former cotton plantation and leper colony in the Caribbean, and turned it into a personal bird sanctuary to house fifty birds of paradise that he shipped from Aru Island, off New Guinea. The birds cost him one thousand pounds, five times more than he paid for the island. But the colony flourished as the jungle gradually reclaimed the plantation over the subsequent half century, and was still going strong in 1963 when a hurricane devastated the small island; the surviving birds were swept out to sea by the winds and drowned.

INTO THE HEART OF DARKNESS

During the eighteenth and early nineteenth centuries, the rainforests of Africa had one great purpose as far as Europeans were concerned – and that was to supply slaves for the plantations of the Americas. The traders in those days had it easy: they rarely had to penetrate the interior and they could rely on local agents and the victors of tribal wars to bring captives to the coast for transportation. Slavery was not new to Africa, but what Europeans did, as they did with rubber, was to industrialize it – with brutal results. Over a couple of centuries, many millions of the continent's poorest and most defenceless – the refugees and prisoners of war, the debtors and the landless – stumbled out of the forests after walking hundreds of miles, often yoked with wooden collars and with the constant imposition of the whip. They were incarcerated among the swampy mosquito-ridden lagoons of the West African coast before being sold to European traders, who shipped them west on a second nightmare journey during which at least a quarter typically died.

By the early nineteenth century, in a rage of moralistic fervour at the height of their global dominance, the British decided to abolish slavery and to use the Royal Navy to impose their new policy on the rest of the world. Britain had no desire to relinquish control of the jungles, however. Far from it. Through the nineteenth century, the triple purposes of enforcing an end to slavery, converting natives to Christianity and making money, saw a huge upsurge in the colonial endeavour. The British – and subsequently the Dutch and French, Germans and Belgians – gave up the simple business of swapping goods at the forest border, and penetrated for the first time deep into the interior. Now they wanted not just to trade, but also to administer and evangelize.

Perhaps the greatest hero of this movement was David Livingstone, a Scottish missionary and explorer who set off north from a small mission station in the Kalahari desert in southern Africa in the 1850s. At first he found vast areas of fertile soils and marshlands and cattle stockades along the Zambezi river; and then, beyond, the beginnings of dense jungle. His purpose was to propagate the Gospel and root out the last slave traders. But he also wanted to record what he saw and to encourage others to follow and 'open up' the continent. He was among the first Europeans to map the continent's jungle heart. As a medical doctor, he noted especially the medicinal plants, many of which he tried out in his efforts to stave off recurring bouts of malaria. And he also recorded local poisons, including extracts from the seeds of *Strophanthus*, a climber in the canopy of rainforest, that could, he noted, stop a buffalo in its tracks.

In his many writings and speaking engagements, Livingstone tried to present an upbeat image of Africa as a land of evangelical and economic opportunity. But he also had another message, for Livingstone had a very modern fear that the African environment was going to the dogs. As early as 1843, he wrote from southern Africa to his masters at the London Missionary Society that 'one of the finest watered countries in the world' was being turned to 'sterile waste'. Many decades before modern environmentalists invented the word desertification, he was arguing that something of the sort was under way.

Livingstone believed that a great and permanent desiccation was occurring almost everywhere he went, including, improbably, deep in the rainforest. Again like modern 'greens', he believed that the destruction of forests was probably behind the desiccation. His warnings led to a minor panic among colonial administrators and scientists. At a meeting at the Royal Geographical Society in 1865, naturalist James Wilson said that the Kalahari desert was spreading as a result of the 'reckless felling of timber'; and Francis Galton, a famous scientist of the day and cousin of Charles Darwin, blamed the felling on the sale of cheap axes to the natives. The panic subsided, however, and the desiccation turned out to be the simple result of a natural cyclical drought. But the seeds of near-permanent concern for the drying out of Africa were sown.

During Livingstone's triumphant returns to London in 1856 and 1863, it was inevitably his tales of danger, disease and derring-do that were most eagerly lapped up by the Victorians in their drawing-rooms. This, they were determined to know, was a dangerous land of pestilence – 'a welter of putrefaction, where men die like flies'. And Livingstone's fame excited others: if a modest and solitary Scottish doctor could almost by accident become a national hero, then others, with sometimes baser motives, sought to join in.

These included a belligerent Welshman called Henry Morton Stanley. Born 'John Rowlands, bastard', he was a former inmate of a Welsh workhouse. He fled to the US as a cabin boy, fought on both sides in the US Civil War and became a journalist. He travelled the world for the *New York Herald* and its proprietor James Gordon Bennett and, in 1871, in that role, he sailed for Africa to find Livingstone. (Though nobody had heard from him for a while, the poor chap was not lost but merely searching for the source of the Nile.) Stanley's expedition, culminating in his discovery of the doctor on the shores of Lake Tanganyika, resulted, of course, in the brilliant piece of journalistic self-quotation: 'Dr Livingstone, I presume.'

While on Livingstone's trail, Stanley filed eight months' worth of dispatches that borrowed from and further refined the great narrative of European explorers in Africa. As Adam Hochschild put it in his own, very different, epic, *King Leopold's Ghost*, 'there were the months of arduous marching, the terrible swamps, the evil "Arab" slave traders, the mysterious deadly diseases, the perilous attacks by crocodiles, and finally Stanley's triumphant discovery of the gentle Dr Livingstone.'

Stanley claimed once to 'detest the land' of Africa. He wrote after one particularly bad day of 'a murderous world . . . we hate the filthy, vulturous ghouls who inhabit it'. He was notorious even in his day for his brutality towards the men employed to guide him across the continent and for launching vicious attacks on native villages in his path. He wrote in his diary how a 'dog-whip became the backs' of his bearers, how his porters were 'faithless, lying, thievish, indolent knaves, who only teach a man to despise himself for his folly in attempting a grand work with such miserable slaves', and how those who tried to flee were 'well-flogged and chained'. If we know of Africa today as the 'dark' continent, it is largely because of Stanley's frequent use of the word, especially in the titles of his widely read books *In Darkest Africa*, *My Dark Companions*, and so on.

* * *

Why do it? Why go there? Fame was part of it, and the urge for derring-do. But for Stanley, as for many others who went, the heart of Africa was an 'unpeopled country, one wide, enormous blank' upon which he could make his mark. He imagined one day 'a church spire rising where that tamarind rears its dark crown of foliage, and think how well a score or two of pretty cottages would look instead of those thorn clumps and gum trees.' But amid the bucolic imaginings, there was also money to be made. As Theodore Roosevelt put it a few years later, of the same continent: 'Surely such a rich and fertile land cannot be permitted to remain idle . . . The country along this [Congo] river is fine natural cattle country, and one day it will see a great development.'

Another explorer to argue this was Verney Lovett Cameron. A young Scottish naval officer, bored with his time at the steam reserve in Sheerness on the Thames Estuary, Cameron volunteered in the year after Livingstone's 'discovery' by Stanley, to do what Stanley had failed to do – persuade the great man to come home. But it was too late. After many travails, Cameron was travelling slowly west from Zanzibar towards the great lakes when he came upon Livingstone coming in the other direction – embalmed, in a coffin, and on his way back to London, where he was finally buried in Westminster Abbey after a state funeral.

Thwarted in his original aim, Cameron seized his moment and embarked on his adventure, heading west into the Congo basin. Two years later, in 1875, he arrived in modern-day Angola, having completed the first east–west journey across Africa by a European. After that he told *The Times* that 'vast fortunes will reward those who may be pioneers of commerce . . . the interior is a mostly magnificent and healthy country of unspeakable richness. I am confident that with a wise and liberal (not lavish) expenditure on capital . . . nutmeg, coffee, groundnuts, oil palms, rice, wheat, cotton, India-rubber, sugar cane . . . may be made profitable. A great company would have Africa open in about three years, if worked properly.'

The British proved none too interested in Cameron's agricultural ambitions, but, within months, King Leopold II of Belgium asked to meet him. Leopold had his own ideas of African conquest and wanted to enlist Cameron in setting up a personal empire in the heart of the Congo basin, which had not yet been claimed by any colonial power. But Cameron hummed and hawed and his fellow Livingstone-chaser, the brusque and bruising Stanley, stepped in. To stake his claim, he too began a journey across the Congo basin. Two years after Cameron completed his own crossing, Stanley emerged from the jungle at Boma, a European station near the mouth of the Congo river.

Stanley had traversed 1,600 kilometres of what he saw as one of the last great empty spaces on the map of Africa, with one of the largest expeditionary forces ever to descend on the continent – 356 people, in all. Moreover, unlike Cameron, he had done it by following the great river. Arguably that was the soft option, but it was more useful to Leopold, since it traced for the first time a route into the heart of the second greatest rainforest on the planet – a route by which colonization could take place. The journalist was already fêted as the intrepid explorer, and was now about to become the intrepid entrepreneur. Over the grave of Livingstone, the great missionary and explorer, the real business of conquering the African interior for commerce began.

* * *

In 1878, as Stanley stepped back on to European shores, Leopold's agent was on the dockside waiting to hire him to head back to the Congo. The journey was to be under the guise of an international expedition of exploration. But, in reality, it had the explicit intention of carving out the first great European commercial operation in the African jungle – initially to harvest elephant ivory. It turned into what a Polish-born master mariner called Konrad Korzeniowski (who wrote about his journey to the region in 1891 under the name Joseph Conrad) called 'the vilest scramble for loot that ever disfigured the history of human conscience'.

In the last years of the nineteenth century, most of the Central African rainforest was annexed by Leopold as his personal fiefdom. This was done in large part through the endeavours of Stanley who, in five years of employment with

Leopold, pressed four hundred African tribal chiefs to sign pieces of paper that (though probably few knew it at the time) gave over sovereignty of their lands to Leopold. In 1885, Leopold declared himself the King-Sovereign of the Congo Free State, and from then on, he, Stanley and the European companies to whom they leased large tracts of the jungle, did what they liked, regardless even of the wishes of the Belgian government. Leopold had no time for setting up plantations or engaging in educational or even missionary work: he was there to make money from ivory; and, regarding the locals as 'a race composed of cannibals for thousands of years', he had little compunction about how he did it.

This most unfree of free states was seventy-five times the size of Belgium and, in Leopold's name, thugs roamed the land on ivory raids, shooting elephants and confiscating tusks as they found them. The agents of the King-Sovereign were on commissions and, when ivory was no longer lying around, they rounded up hundreds of thousands of Africans and sent them into the jungle to kill more elephants. The ivory was shipped downstream and away to Europe, where it made piano keys and billiard balls. When the gangs failed to deliver enough ivory, use of the chicotte – a whip made of sun-dried hippopotamus hide cut into long, sharp-edged strips – was widespread. A hundred lashes was usually fatal.

Stories of the brutal beatings and torture inflicted by Leopold's personal army, the *Force Publique*, seeped back to Belgium, but few cared. Meanwhile, Stanley's involvement in all this stood uneasily with his public attitude to what was going on in his new land. He wrote in his book *In Darkest Africa* how 'every tusk, piece and scrap of ivory . . . has been steeped in human blood. Every pound weight has cost the life of a man, woman or child; for every five pounds a hut has been burned; for every two tusks a whole village has been destroyed. It is simply incredible that, because ivory is required for ornaments or billiard games, the rich heart of Africa should be laid waste and that native populations, tribes and nations should be utterly destroyed'. That was a bit rich from the man who did more than anyone to establish the trade and the terror.

While emissaries of the King of Belgium ruled the Congo with a heart of darkness, most of the ivory ended up in Britain, which imported an average of 500 tonnes a year – half of total world consumption. Much of the rest went to the US, where, at the turn of the century, 200 tonnes of ivory was being used annually for making keyboards so that the well-to-do could 'tickle the ivories' of the newly popular upright pianos. The world consumption of ivory at the time represented the deaths of sixty-five thousand African elephants a year, a scale of slaughter that was decimating the population of the planet's greatest land mammal.

* * *

Stanley did not start the ivory trade, it is true. A decade before, Cameron had recorded its impact during his journey across the Congo basin in search of natural resources to harvest. He noted that 'ivory is not likely to last for ever (or for long) as the main export from Africa; indeed the ruthless manner in which the elephants are destroyed and harassed has already begun to show its effects. In places where elephants were by no means uncommon a few years ago, they are now rarely encountered.' Having this probable extinction of the ivory trade in view, Stanley proposed harvesting vegetable and mineral products instead. And he was proved ultimately right. After a decade of ruthless plundering of the continent's elephants, Leopold moved on to a new crop. This was a widely found vine of the *Landolphia* genus, which produced a rubber latex similar to that which had proved so profitable in the Amazon. The plant and its latex had first been spotted by Pierre Poivre, the enterprising French clove smuggler from Mauritius, but it had not previously been commercially exploited. Seeing the Amazon rubber boom in full spate, Leopold launched his own version, which soon became known as the 'rubber terror'.

Agents once again swarmed across Leopold's 'free state' demanding the extraction of natural rubber from the forest. By the turn of the century, exports of rubber had reached 6,000 tonnes a year. This was sixty times what it had been a decade before, though less than a quarter of what was being exported from the Amazon at the same date. But whereas Amazon rubber tappers could take the sap without killing the tree, the African vines were ripped up in an orgy of production that wrecked the forest across huge areas of the country.

As the vines disappeared from sites near their villages, Africans were forced to travel ever further from their homes to find it – and as their harvests diminished, the brutality intensified. Whole villages were taken hostage, their women raped and their livestock taken, until the men produced the required amount of rubber. The local British consul Roger Casement, who was appalled by what he saw, reported back to London on the depopulation of whole areas of the country. He quoted native rubber collectors who told him: 'We tried, always going further into the forest, and when we failed and our rubber was short, the soldiers came to our towns and killed us. Many were shot, some had their ears cut off; others were tied up with ropes around their necks and taken away.' (Casement later played a part in exposing the rubber terror of Julio Cesar Arana in the Amazon.)

Soldiers also cut off the hands of those they killed. 'These hands – the hands of men, women and children – were placed in rows before the Commissary, who counted them to see that the soldiers had not wasted cartridges,' wrote an investigator and campaigner against the slaughter, a young English journalist called Edmund Morel. Some estimates put the number killed during this reign of terror at a staggering 8 million. It ended only when the Belgian government took over the fiefdom in 1908. By then the human population of the Congo basin was halved. As John Reader, in his biography of Africa, puts it archly: 'Leopold displayed exceptional generosity in the disbursement of his new-found wealth. The Congo profits were used to fund a grandiose policy of public works and urban improvements – in Belgium.' (Not for him an opera house in the heart of the forest.)

The era is summed up by Conrad in his novel *Heart of Darkness*. Narrator Charlie Marlow makes a fictional journey up the Congo river that mirrored the journey Conrad himself had made. He describes vividly how 'going up that river was like travelling back to the earliest beginnings of the world, when vegetation rioted on the earth, and the big trees were kings. An empty stream, a great silence, an impenetrable forest. The air was warm, thick, heavy and sluggish . . . It was the stillness of an implacable force brooding over an inscrutable intention. It looked at you with a vengeful aspect.' It was a country in which, in the time of Leopold, 'the word ivory rang in the air, was whispered, was sighed. You would think they were praying for it.'

And upstream, at the end of the book, Marlow finds a white ivory trader, Kurtz, who has amassed huge quantities of the stuff on his remote station at the head of the river Congo. He is found sunk into savagery and dreaming megalomaniac dreams, surrounded by ivory and human heads stuck on to fence posts. The book asks rhetorically whether the 'heart of darkness' is in Africa or in the white invader. Did Kurtz (whose dying words are 'The horror. The horror!') find evil in the rainforest? Or did he corrupt the forest with his own evil? It has remained a potent question ever since. And not just in Africa.

REPLANTING RUBBER

By the 1890s, the rubber booms in Brazil and the Congo were already doomed. The explosion of activity that some believed would propel Brazil into becoming the world's first industrial tropical jungle state was about to end. Not content with grabbing control of the rubber trade in South America and removing its processing to their own territories, the industrialized countries wanted to assume control of the growing of rubber as well and to industrialize it. And they didn't want to do that in Brazil.

The botanists at Kew in Britain already had a growing reputation as a clearing house for the globalization of commercial plants. Among a large collection of plants sent back from the Amazon by Spruce in 1854 were some rubber trees. And in 1870, fresh from bringing cinchona seeds to Kew, Clements Markham recommended to his bosses that it was 'necessary to do for the India-rubber-yielding trees what has already been done for the cinchona trees'. Why necessary? Clearly imperial advance was involved – but, as with cinchona, the Kew crew argued that it was an environmental necessity, too. There were dire warnings that 'unless the natives can be prevented from destroying the trees', they would be driven to extinction.

Joseph Hooker, the then director of Kew, backed the idea and chose *Hevea brasiliensis* as the best among the samples sent back by Spruce. Searching for a suitable adventurer to go and capture some seeds, he wrote to Henry Wickham, a coffee planter and former trader in exotic bird plumage, who had recently offered his services. The handlebar-moustached Wickham was a bit of a card and, like Markham, a great embellisher of a story. His tale of how he 'smuggled' rubber plants out of the Amazon, pursued by a Brazilian gunboat, has become a seminal tale of botanical colonialism. Both Victorian colonialists and their modern critics seem to relish the story, as either heroic endeavour or dastardly piracy.

In truth it was neither. For one thing, there was no great subterfuge. Sometime in June 1876, Wickham simply presented his specimen seeds to Customs at Belem in a box marked 'exceedingly delicate botanical specimens specially designated for delivery to Her Britannic Majesty's own Royal Garden at Kew'. They were waved through. And for another, there was, in fact, no Brazilian law prohibiting the export of rubber seeds. Such trade was widespread, and several other collectors had already removed rubber plants from Brazil at Hooker's request and without trouble from Customs. They included stalwart Kew collector Robert Cross, who had made two journeys, surviving shipwreck on one occasion, and was headed back to the Americas for a third time before Wickham, who had haggled with Kew for months over his fee, made his first delivery. But whereas Cross's mixture of seeds and cuttings proved weak and useless, Wickham's box of seeds, planted out in the grounds of Kew the day after they arrived on a special train from Liverpool, germinated and proved hardy. As a result, they were successfully transferred to Ceylon and planted out during the 1877 monsoon.

It takes time to establish rubber plantations, and

during the coming years British imperial endeavours were equally involved in finding native rubber in Africa – where they succeeded both along the Zambezi and in Ghana, creating local rubber booms in the last years of the nineteenth century. But by the time Belfast vet John Dunlop came up with his pneumatic tyre, and motor-car manufacture took off, the Far-Eastern rubber plantations were ready to take control of the next rubber boom.

By the early decades of the twentieth century, Malaya and the Dutch East Indies had overtaken Brazil as the leading producers. In Malaya alone, some 40,000 hectares of rubber plantations were dominating the world market and the booms in Brazil and the Congo turned to bust. Brazil tried to establish its own rubber plantations, but yields were poor because an indigenous leaf fungus ran riot through plantations whenever rubber trees were planted close together. There was a brief revival in harvesting wild Brazilian rubber during the Second World War, after the Japanese invaded Malaya and took over most of the world's rubber plantations, but it ended with the war. The many rubber tappers who lived on in the rainforests survived mostly by harvesting Brazil nuts.

TWO MODERN HEROES

The great days of botanical exploration in the rainforests may have been in the Victorian era, but some twentieth-century trail-blazers came pretty close. Take Richard Schultes, who died in Boston as recently as 2001. He was widely known at Harvard as 'the last of the great Victorian plant explorers', and yet he was also the man who invented hippie psychedelia.

Schultes's sense of adventure began early. Lying in bed with a stomach upset at the age of five, he was read to by his parents from the travel diaries of Richard Spruce. This was around 1920, and the diaries were then only a decade old. It was a heady brew, and Spruce's tales captured the boy's imagination. Schultes said later that he decided then and there to follow in Spruce's footsteps. And he was true to his aim, spending many years travelling the Amazon rainforests, and becoming the world's foremost modern authority on the Amazon's medicinal and, especially, hallucinogenic plants.

Drugs were Schultes's thing from the start. While still an undergraduate at Harvard, he wrote a paper on the mind-altering properties of the cactus peyote, after undertaking personal research with the Kiowa Indians in Oklahoma. Did he partake? Of course. 'It would have been unpardonable rudeness to refuse,' he said later. For his doctoral thesis he went to the forests of Oaxaca in Mexico, where he researched the sacred mushroom teonanactl used in rituals by the Mazatec Indians. A female shaman called Maria Sabina told him how 'the more you go inside the world of teonanactl the more things are seen. You see our past and our future . . . I knew and saw God: an immense clock that ticks, the spheres that go slowly around, and inside the stars, the earth, the entire universe, the day and the night, the cry and the smile, the unhappiness and the pain.'

While in the Mexican forests, Schultes also found ololiqui, a vine whose seeds, widely known today as morning glory, contain a natural form of LSD. The Spaniards had known about this drug from their first encounters with the Aztecs – they could hardly miss it. One missionary reported how the Aztecs 'communicate with the devil when intoxicated with ololiqui and are deceived by various hallucinations which they attribute to their deity, which they say resides in the seeds'. What student of the 1960s would not have dreamed of conducting such research? But this was the late Thirties.

Travelling to the Colombian Amazon for the first time in 1941, Schultes went in search of the many forest plants that produce the barks containing the alkaloids used in curare, and catalogued them in a way not attempted since Humboldt's journey 150 years before. He found the substances in more than seventy plants. He travelled by aluminium canoe and wore a pith helmet, but never carried a gun. His heyday was the Forties and Fifties, when he lived almost continuously in the rainforests. So deep was he in the Amazon that it was more than a week before he heard about the Japanese attack on Pearl Harbor. Realizing there was a war on, he presented himself to the US authorities in Bogota for national service, but was sent back into the jungle to revive the long-dormant wild rubber harvest to replace Malayan rubber plantations lost to the Japanese.

In all, Schultes collected more than twenty-four thousand plant specimens from the Amazon, and documented more than two thousand that were used for medicines by the dozens of tribes of Indians whom he befriended. Unlike the conquistadors of old – and many more recent travellers – he maintained the philosophy that 'I do not believe in hostile Indians. All that is required to bring out their gentlemanliness is reciprocal gentlemanliness.' Many claim that Schultes was the inventor of modern ethnobotany – though, as far as he was concerned, he was simply reviving Spruce's habit of tapping local shamans' knowledge. More than a hundred plants that he discovered for the Western world carry his name, including one with a bark whose ashes can treat ulcers, another with a stem that makes a tea that relieves tubercular coughs and a third whose leaves cure conjunctivitis. A chunk of protected Colombian rainforest is named after him.

Schultes liked the glamour of rainforest stories and, in later years, demonstrated the use of poison-dipped darts during lectures. But, like his hero Spruce, he eschewed the false heroics that often accompany jungle travel. He once told the film-maker Herbert Girardet: 'There are too many books by people who go to the Amazon for two or three weeks and escape horrible death on every other page.' His obituaries were instead full of his encounters with the gurus of the drug-fuelled counter-culture. His botanical texts were read by the likes of Aldous Huxley and William Burroughs. One of his papers on hallucinogenic fungi spawned a *Life* magazine article entitled 'Seeking the Magic Mushroom' that inspired Timothy Leary. But Schultes was never seduced by the 'turn on, tune in and drop out' culture of the Sixties. When Burroughs described to him the mind-blowing experience of one psychedelic mushroom 'trip', Schutels reputedly responded drily: 'That's funny, Bill. All I saw was colours.'

* * *

High in the Andean mountains of Ecuador are cloud-covered ridges that no satellite has ever observed and no cartographer has ever mapped. Local rumour has it that, up there somewhere, the Incas hid their treasure when Spanish conquistadors invaded their country five hundred years ago. Treasure-hunters still come looking: so far they have failed to find the hoard, but another prospector has been prowling these hills. Lou Jost, an American scientist and botanical adventurer for the twenty-first century, is another self-confessed disciple of the great Victorian explorer in these hills, Richard Spruce. And he claims to have uncovered the mountains' real treasure – an ecological El Dorado of orchids found nowhere else on the planet.

Jost gave up life as a quantum physicist to take up botany. Since 1997, he has been living in the Ecuadorian Andes, collecting dozens of new orchid species in the remote cloud forests and valleys. He operates alone, without the help of any academic body, but on meagre grants from conservation charities. He explores remote areas shrouded in permanent cloud, above the river Pastaza, a tributary of the Amazon, which carves its way through the Andes and down into the Brazilian rainforest. The valley and the forest-covered mountains that surround it have more endemic orchids than anywhere else on the planet. One of his favourite cloud-covered haunts, the Sacha Llanganates mountains, has never before been seen from a distance, let alone explored on foot. 'This region really is special,' he says.

Jost's tiny greenhouse, housed on the roof of his apartment in the adventure-tourist town of Banos, harbours a collection of local plants found nowhere else. Why? Because, he says, the Pastaza valley is the deepest, straightest valley in the eastern Andes. Every afternoon a hot, wet wind blows up the valley from the steaming Amazon lowlands. It brings huge volumes of moisture that evaporate to form near-permanent cloud and fog that permeate the forests that cover the precipitous mountain ridges. 'Each ridge has its own microclimate in the clouds,' Jost says. 'The first ridge west of the rainforest is the wettest and windiest. The next is slightly less wild and wet. But every one offers a unique environment, and that usually means unique orchids.'

In these wet, sunless environments, dozens of species of tiny, delicate orchids have evolved amid the trees. Most grow on the branches of the trees themselves in tiny suspended ecosystems in the clouds. 'Each species seems to specialize in a particular combination of rain, mist, wind and temperature,' Jost says. 'High in the clouds, you can come across whole areas of forest smothered in a single species of orchid that probably exists nowhere else on Earth. It is an amazing experience.'

Ecuador's cloud forests are planetary hotspots for plants.

Current records show more than four thousand species native to a country the size of Nevada, and the number is being added to all the time. Most live in its cloud forests, and four out of five species are threatened with extinction. The Ecuadorian cloud forests are Cinderella ecosystems, says Philip Bubb, author of the first-ever serious attempt to map them. He says that there are many similar ecosystems in highlands of tropical islands like New Guinea that have never been explored by biologists. 'Cloud forests are fantastically beautiful and lush,' he says, 'with orchids, mosses and ferns growing across every surface. Each branch is like a garden in itself. And the atmosphere is damp and cool with an eerie mist in which bird calls carry vast distances.' These are genuinely unexplored regions of the Earth. Even large animals survive undetected: in the cloud forests of Laos and Vietnam, scientists have discovered two new species of deer within the past decade.

The Pastaza valley is the heartland of that diversity in Ecuador, and Jost has identified ninety orchids unique to the valley during his six years' study. On one red-letter day alone, he found four new species of Teagueia orchids – one purple, one black with antelope-like horns, one orange and one striped tan and purple. All were hidden in a single patch of moss on Mount Mayordomo. That one day's botanical prospecting raised the number of known species in the Teagueia family from six to ten. And since then, he has found another sixteen long creeping Teagueia orchids on the mountain.

In recent years, Jost and other botanists have found 197 unique plant species in the Pastaza valley – more than the

180 found on Ecuador's other, better-known, biological treasure house, the Galapagos Islands. 'The Galapagos is fully studied – pretty much every booby fart there is recorded – but up here we have huge areas that have never been explored,' he says with relish. 'The only way to discover the botanical secrets of this area is to walk every ridge and valley. And you have to know what you are looking for. The flowers are only a few millimetres across and usually hide under the leaves. Often the plants are not in flower. If I spot what I think is a new species I can often only be sure when I bring it back to my greenhouse to wait for the flower to appear. Only then do I know if I have a new species.'

Jost believes that his findings conflict with conventional thinking about the evolution of endemic plants. The usual view is that endemism is caused by geographical isolation: the plants could grow elsewhere, but they cannot escape the confines of their single habitat. 'That's not true for these orchids,' he says. 'They have tiny dust-like seeds that can spread easily. But the colonizations fail because they literally cannot grow anywhere else.' No cloud forest is like any other – and for orchids that makes all the difference. 'Subtle climate variables are the driving force behind the evolution and survival of endemic species among these orchids.'

Jost is a botanical Indiana Jones. Despite being on the equator, it is bone-chillingly cold and wet up in the mountains, and the area is riddled with cliffs and impassable ravines that are all the more dangerous in the near-permanent mist. What trails there are have mostly been made by the mountain tapirs. Some parts are guarded by tenacious Shuar Indians, perhaps descendants of the Inca gold-hoarders. Most are empty, except for the occasional tapir and spectacled bear. But whether braving bears, frost-bite or independent-minded locals, Jost rejoices in following in the footsteps of Spruce, who trekked through the Pastaza valley in the 1850s after crossing the Amazon basin. Spruce did things and went to places that no botanist has done since, says Jost. He discovered ferns and liverworts in the Pastaza valley that nobody has seen since. And in the twenty-first century there are still such finds to be made.

LAWS OF THE JUNGLE

Nowhere are so many species mingled together, with such elaborate inter-connections, as in the rainforests. Even forest giants like the Brazil-nut tree and forest killers like the strangler fig depend for survival on cohorts of helpers on the jungle floor. How does this exquisite co-operative endeavour fit with Darwin's theories of survival of the fittest? Or, come to that, with new ideas about the randomness of the rainforest? How little we know. We are still discovering new species of the largest mammals, still looking for mythological beasts and still being hoaxed by locals with a twisted sense of humour. And there remains a huge and largely undiscovered world high up in the forest canopy, where forest life is at its most abundant, where the forest controls the very air we breathe – but where, to this day, few humans dare go.

WHAT MAKES A RAINFOREST?

The tropical forests have their origins almost 200 million years ago, during the early days of the dinosaurs. The continents had not yet separated: they were huddled together in a giant, supercontinent called Pangea. The rest of the planet formed a single giant ocean, today's Pacific Ocean, but then even larger. Pangea was centred around the equator and occupied by huge ferns and early forms of conifer trees. They would have felt rather like today's hot, sticky rainforests, only hotter and stickier. But they still lacked the flowering plants that gradually took over in the ensuing 100 million years, and which still dominate most forests today.

At the same time as the modern forest was evolving, Pangea was gradually breaking up. First Australia and Antarctica (which was once itself forested) broke away. Then the Americas began to move off, creating the Atlantic Ocean. Later, Madagascar separated from Africa. At each stage, evolution started to take a different course, forming different branches on each continent. Thus the forest monkeys that evolved in the old and new worlds, while coming from the same ancient stock and living similar lives in similar environments, are anatomically now quite distinct. Similarly, lemurs survived only in Madagascar, partly because the island had no large predators.

Since the days of Pangea, the planet's climate has cooled somewhat; but throughout the wet tropics, where it rains most of the year, rainforests remain the natural vegetation. By far the largest of these forests is in the Amazon basin of South America. It contains roughly two-thirds of the world's surviving tropical rainforests, representing some 30 per cent of all the biological material on the surface of the planet. The forest receives so much rain that the river running through it has five times the flow of the next biggest river, the Congo, which runs through the second largest continuous tract of rainforest, in Central Africa. The third great rainforest region straddles southeast Asia from Burma through Malaysia and Indonesia and on to the islands of the South Pacific, such as New Guinea and the Solomon Islands. Smaller patches of rainforest also survive in West Africa, Central America, northern Australia, the Indian subcontinent and on some tropical islands – but most of the forests in these areas have been lost to economic development.

The great mass of rainforests are on low land, in the basins of great rivers like the Amazon and the Orinoco in South America, the Congo in Central Africa and the Mekong and Irrawaddy in mainland southeast Asia. Most are on dry land, creating the dark and surprisingly vegetation-free 'cathedrals' beneath the canopy that many rainforest explorers discuss in awe. But some occupy land that is seasonally inundated by the rivers, creating huge zones of flooded forests, such as those in the central Amazon, or sit on top of the great peat swamps of Borneo and Sumatra in Indonesia. Other rainforests form on coasts where dense thickets of mangroves still thrive, their distinctive stilt-like roots probing the water to find soil. The biggest of these are the Sunderbans mangrove forests on the Ganges delta of India and Bangladesh – the famed domain of the Bengal tiger – and on New Guinea, the world's largest tropical island.

The final rainforest category is the cloud forests of the tropical uplands. These grow in near-permanent clouds on the eastern slopes of the Andes as they slope down into the Amazon basin, in the highlands of Central Africa, in parts of Central America and in the remote interior of New Guinea. Cloud forests are generally the most remote and least explored. The trunks of the trees are short and gnarled by comparison with the tall straight trees of the lowlands, but their canopies are among the richest of all, with immense growths of ferns, orchids, lichens and mosses luxuriating in the damp of the clouds.

* * *

For biologists the most dramatic thing about the rainforests is the vast number of species they contain. Though they occupy less than 7 per cent of the planet's land surface, they contain at least half, and perhaps as much as two-thirds, of all the species that live on the planet. A third of all the world's nine thousand known bird species live in the Amazon alone. Even small areas of forest contain great diversity. A single hectare of Malaysian rainforest – about the size of a soccer pitch – typically contains 180 different tree species. A similar patch in Peru may contain three hundred tree species. One entomologist, exploring a small plot of rainforest in the Brazilian state of Rondonia, counted 429 butterfly species in twelve hours. Another found eighteen thousand species of beetles in a single hectare of Panamanian forest.

There are many theories about why this huge biodiversity should occur, and many answers to the question first posed by Charles Darwin a century and a half ago, after his first journey to the Brazilian rainforests: 'What explains the riot?' One theory is that it is the unchanging, primeval nature of rainforests that has caused this biological cornucopia. As Paul Richards, the British author of a classical text, *The Tropical Rainforest*, put it: 'The immense floristic riches of the tropical rainforest are no doubt largely due to its great antiquity; it has been the focus of plant evolution for an extremely long time.' Some have coupled that antiquity with the high inputs of energy from the sun that can push evolution into overdrive as hordes of plants and creatures rush around trying to maximise their potential in the sun. Maximum biodiversity, in other words, requires plenty of energy, as well as plenty of stability.

But others dismiss this beguiling image of a primeval cornucopia. They say that it may, paradoxically, be the very temporary nature of many forest environments that breeds diversity. Disturbance breeds diversity. One strand of this argument concerns what happened to the rainforests during the ice ages that have punctuated the climatic history of our planet over the past 2 million years. The ice caps never reached the lowland tropics but, as they spread towards the equator, the tropics both cooled and became drier. Back in the 1970s, a German ornithologist called Jurgen Haffer and British botanist Ghillian Prance first argued that, during those times, the rainforests retreated into core areas for a few tens of thousands of years and grasslands invaded their former domains. There would have been extinctions as the forests drew back, they agree, but the most adaptable species would have hunkered down in these ice-age Noah's arks, waiting for the climate to become warm and wet again.

The core forest areas that survived through the cold eras, often called refugia, would have become forcing grounds for evolution. In them, so the argument goes, new species would have emerged to take advantage of the changed circumstances. And in each refugia, evolution would have formed new branches, much as it did when the continents separated. When the good times returned, the refugia disgorged their new species across the wider rainforest. The evidence for this argument has come mostly from the Amazon basin where, to this day, the greatest biodiversity is concentrated in hotspots that – according to the theory – were the forest island refugia during the ice ages. One such is the western Amazon region where Colombia, Brazil and Peru meet. But the forests may also have retreated over large areas of Central and West Africa during the ice ages, and probably in the Far East too.

This is still controversial territory. Some researchers see no evidence for such a dieback of the forests. They believe that most of the Amazon rainforest survived the more recent ice ages, though the mix of species could have changed dramatically to the kind of dry forest species now seen only in small remnants in eastern Brazil. In any case, they say, whether refugia formed or not, they don't explain the fantastic biological diversity of rainforests. Evolution wouldn't have happened that fast. And there is no obvious reason why the gains should have so dramatically outweighed the losses as the forests retreated.

But there is a second strand to the argument, that change rather than stability created the rainforests' stupendous biodiversity. This is about dynamic, day-to-day change rather than the long-term waxing and waning of the ice ages. Roger Lewin of the journal *Science* summed up the vision: 'Tropical forests are dynamic and ephemeral rather than stable and ancient . . . High diversity through instability, not stability, is how the equation now reads.'

Far from being unchanging, many rainforests are prey to dramatic natural disturbances like landslides, hurricanes and fires that can destroy huge areas. While rainforests are usually too wet to burn, whenever droughts occur – as across the Indonesian rainforests in 1998 during the El Niño event there – they burn with a vengeance. These disturbances change everything. The dark, empty forest floor is suddenly drenched in sunlight and a burst of plant growth occurs, until the new trees grow high enough and the canopy closes again some decades later. Every time this happens, new species invade and evolution gets another push. As the light gap is filled, the new species mingle with other species that prefer things darker to create a complex mosaic of habitats that maximize the living space for innumerable species and generate biodiversity.

This theory really shakes the certainties of environmentalists because, since the ending of the last ice age, it may well have been humans, as much as hurricanes or landslides, that have been the major cause of disturbance in the rainforest. That opens up the idea that maybe we, too, have a hand in keeping this evolutionary engine on the road. Biologists are coming to realize that many of the most biodiverse rainforests are also the most disturbed by humans.

* * *

But more of that later. For there is now a third idea doing the rounds that is really upsetting the apple cart. It turns the ideas of Darwin and many others on their head. The law of the jungle, says Steve Hubbell, an iconoclastic plant ecologist from the University of Georgia, is not about the survival of the fittest at all. It is about the survival of the luckiest. Hubbell has spent two decades on Barro Colorado island, a natural rainforest 'laboratory' in Panama run by the Smithsonian Institution (see p.120), so he can claim to have examined the dynamics of rainforests more closely and for longer than most. He has concluded that many, perhaps most, species survive more by luck than through Darwinian selection pressures. In some instances, he says, it is not the strongest that survive, but the weakest. And far from fighting pitched battles for supremacy, jungle species may get by just because there is very little competition.

Take Hubbell's study of how young plants fill the gaps in the rainforests when mature trees die. According to the disturbance theory of rainforest biodiversity, it is this moment, when a giant tree crashes to the ground, allowing light and rain to stream to the forest floor, that is the great driving force of evolution and biodiversity in the forest. But, after fifteen years spent studying more than 300,000 trees on a 50-hectare plot on the Panamanian island, Hubbell thinks differently. He found no evidence that tree falls and other disturbances boosted diversity at all. The light-splashed gaps created by falling trees contained the same mixture of species as dark, untouched forest. And patches of forest that had been disturbed most in the past contained no more species than the rest.

How come? Well, it seems that our ideas about a highly competitive rainforest environment, with species scrabbling for space and ascendancy, turn out to be rather fanciful. Hubbell found that there was no huge competition to fill the gaps left by the fallen tree. There was no evidence of birds and bats rushing in to drop seeds and start a rush to populate the clearing. Rather the reverse. After collecting more than a million seeds in traps set right across the jungle, he concluded that most trees were lousy dispersers. The jungle was actually short of seeds from most species. Far from tree species being highly competitive, elbowing their way into the clearings to gain a foothold among the thrusting new plants, most seemed complacent about their place in the jungle.

'The gaps are being occupied largely at random,' says Hubbell. 'Many of the sites are occupied, not by the best competitor for the site, but just by whoever happened to be there.' Apparently weaker species thus did as well as stronger ones, because there was no competition.

The apparent losers and ill-adapted types were getting by better than expected. The place resembled a particularly liberal and non-judgemental kindergarten in which 'everyone can succeed', rather than the competitive 'jungle' in which the underachievers are ruthlessly cast aside. Trees and other plants followed the pattern, and so, too, did bats and birds, and fish and frogs. And, he argues, the kindergarten promotes biodiversity just as effectively as a competitive environment. Hubbell's census of trees on the island revealed that there were many more rare species than would be predicted by classical ecology based on fierce competition. The survival of the weakest promotes biodiversity just as effectively, perhaps more so, as the survival of the fittest. 'Live and let live' was the jungle motto in his neck of the woods.

Is there anything special about Barro Colorado that creates this liberal environment? There is no reason to suppose so. To test his observations in Panama, Hubbell built a computer model of evolution in the rainforest, in which there was no chemistry, no biology, no climate – just a range of species experiencing entirely random births and deaths with no evolutionary and environmental pressures on them at all. The only constraint was space. In his simple simulation, a new tree could grow only when another died. With that single constraint, he installed a random number generator into his model to kill and create trees of different species at will. It didn't matter if they were well adapted or ill adapted to the forest: if they could survive, they would prosper.

Hubbell sat back to watch what happened in this randomized rainforest world. Some researchers predicted a rapid ecological breakdown as the conventional ecological rules of the jungle were subverted. But instead things carried on quite happily. 'I got patterns that looked very similar to the patterns I was seeing in the forest,' Hubbell says. The model is only a mathematical construction, of course, but it suggests that tree species do not need to adapt in any complex way to fit their environment, or evolve too hard to compete with their neighbours. They can be determinedly eccentric and suboptimal, sublimely uncompetitive and self-indulgent – and do just as well as their driven neighbours.

This is, of course, a highly political view of the rainforest to which some of us are drawn while others are repelled. Put into human terms, it means that just as a child can learn and prosper in a non-competitive environment as much as in a hot-house school, so a species can do much better than anyone supposed simply by getting along. The rainforest looks much more like the real doing-just-enough-to-get-by world than the pressurized academic world of Harvard or Oxford. This is good news for every slob, underachiever and couch potato on the planet; but for the high-flying scientists researching the rainforest it must have come as a bad shock – they expect rainforest species, in particular, to be highly trained specialists like themselves.

Not surprisingly, Hubbell's grandly titled book encapsulating his ideas, *The Unified Neutral Theory of Biodiversity and Biogeography*, published in 2001, is highly controversial. 'A lot of people would like to call him a crackpot and ignore him,' his co-scientist in Panama, Doug Erwin, told *New Scientist* magazine. Certainly plenty of researchers right now are out in the jungle doing research aimed at knocking down his theories. That is the reality of the academic jungle. But Edward Wilson, esteemed Harvard biologist and popular writer on rainforests – many of whose ideas are undermined by Hubbell's work – takes a more relaxed view. He told the magazine that he expects Hubbell to rank one day as one of the most important ecologists of the past century.

THE PARABLE OF THE BRAZIL NUT AND THE AGOUTI

Whatever the role of randomness in rainforest ecology and the 'survival of the weakest' in evolutionary theory, there can be no doubt that some vital alliances are forged in the jungle. Many species depend on each other in bizarre and unexpected ways. However much we think we know about the forests, it is clear that we are just scratching the surface of these relationships. One such concerns the giant Brazil-nut tree and a forest-floor guinea pig. Peruvian botanist Enrique

THE SMITHSONIANS

The Smithsonian Institution's living rainforest laboratory on Barro Colorado island in Panama is a haven for rainforest research, a constant source of enlightenment about the bizarre jungle world, and easily the most studied stretch of rainforest in the world. The island was created during the construction of the Panama Canal a century ago. The 1,600-hectare island, plus an area twice as large on surrounding peninsulas, has been protected for more than eighty years and studied by Smithsonian scientists for most of that time.

Its trees are logged and measured ceaselessly, its canopy is explored from a giant crane and its clearings are penetrated with radar. Animals like ocelots and agoutis, lizards and antbirds, are tagged with radio transmitters and tracked every second of the day from seven giant radio antennae above the island. The transmitters reveal not just where the animals are, but what they are doing, how fast their hearts are beating and a host of medical details. In one dramatic episode of the Barro Colorado soap opera, an agouti suddenly stops moving and its heart rate drops to nothing. Looking at their computer screens, the Big Brother team can see that an ocelot has stopped at the same spot. After a burst of pumping heart activity, it too stops moving, stuffed and happy.

It was here that Stephen Hubbell tracked 300,000 trees and collected more than a million seeds to show how lousy seed dispersal was at furthering the survival of the fittest and to explore his belief in the true random nature of rainforest evolution. Here, too, that biologist Jae

Choe of Harvard University discovered bats that made wigwams to keep off the rain. Bats of the *Uroderma bio-batum* species gnaw part-way through the leaves of their favourite trees so that the leaves droop under their weight and overlap to form several tiers that keep out the rain. The architecture is sophisticated: the bats know that by chewing about 10 centimetres from the stem, they create the right-sized wigwam, with room to roost and fly in and out. But further up, they chew closer to the stem to form snug extra layers.

It was here that someone noticed moths that will dive on you if you jangle a bunch of keys. And here, too, that Douglas Altshuler studied fruits from fifty-seven plant species to find that the birds that disperse their seeds preferred those that reflected the most ultra-violet light, irrespective of their colour. Birds, it seems, can see whole spectra of light denied to humans. This is a useful as well as a clever trick, since it turns out that as fruit ripens it reflects more ultraviolet light, ensuring that birds get the tastiest fruit and do not waste their time dispersing seeds before they are ready to germinate. To double-check what was going on, Altshuler put UV filters over some fruit and, ripe or not, the birds left it alone. And it was here that John Nason discovered the extraordinary power of mature strangler fig trees to attract the tiny wasps that take their seeds and pollinate new trees. Using genetic markers, he found that the strangler figs in the forest were able to draw the wasps from an area of between 100 and 600 square kilometres.

Ortiz has devoted his life to learning about the Brazil-nut tree, the forests that the tree made and the creatures that made the tree. What he has learned tells us a huge amount about the Amazon rainforest, how species are interconnected in the most unlikely ways, how vital rainforest ecosystems are to the survival of even a tree as dominant as the Brazil-nut tree – and how this ecological house of cards could be brought down.

The Brazil-nut tree is the Godfather of the jungle. It grows so high it soars above the canopy. And it lives for a thousand years, becoming the centre of a vast web of life. Trees growing in the Amazon today were fully grown when the conquistadors arrived half a millennium ago. They saw great Indian jungle empires fall and the rubber boom come and go. Their rings hold records of climate in the Middle Ages and their gaseous emissions are measured by modern satellites circling the Earth.

And their story begins with a nut. The Brazil nut is the hardest nut in the forest. Hard as a rock and heavy as a cannon-ball, with twenty or thirty woody seeds encased inside. Falling from the branches of a tree as high as a church steeple, any one of the coconut-sized pods can kill. And sometimes they do, says Ortiz, whose academic home is the Smithsonian Institution's National Museum of Natural History in Washington DC. 'You sometimes see animals staggering around with large welts where they've been struck.'

Ortiz does his research into the Brazil-nut tree from a thatched hut by a river bank on an old rubber-tappers' reservation in a remote Amazon valley on the border between Peru, Bolivia and Brazil. The Incas called this place Serpent

Valley after the anacondas they found there; today it is called Madre de Dios – and it is part of one of the most biodiverse regions on Earth. Beside anacondas, it is home to jaguars, pumas, eagles, caymans and an extraordinarily productive area of economically valuable plants. It is the land of cinchona and rubber.

Ortiz, a bearded and intense man in his mid-forties, counted and followed the fate of almost every pod that fell on to the forest floor from around a thousand trees. The seeds of the Brazil-nut tree, arranged inside their pods like the segments of an orange, have many hurdles to cross before they can produce a new tree. First, they have to escape their armour-plated pods. He watched and waited as cat-sized, strong-toothed and big-jawed guinea pigs came by. These guinea pigs, called agoutis, chisel their way into the pods and hide the nuts in the ground like squirrels with acorns, keeping the nuts out of the way of porcupines and other rivals that will eat them if they can find them. To keep track of the nuts during his research, Ortiz then dug up some three thousand nuts and attached magnetic strips to each of them, before burying them again for the agouti to collect. Later, he searched the forest with magnets to find out what the agoutis did with them. In this way he showed that the agoutis will cart the nuts up to half a kilometre away. They will return to eat some later, some will be stolen by other creatures, but just one or two may germinate and form new Brazil-nut trees.

Ortiz knows his agoutis as well as he knows his trees, for without the agouti, he says, the Brazil nut would be no more.

While agoutis eat most of the nuts, only the agouti among the forest's current dwellers can open the pods and release the tree's precious seed. None of the other bigger and tougher inhabitants of this forest can manage this trick. Maybe the agouti would get along without the giant Brazil-nut tree; but, assuredly, the tree would die without the agouti scuttling along the forest floor and releasing its seeds. Brazil-nut trees are found only where there are agoutis.

But the mutual support system built around the Brazil nut involves much more than the sharp-toothed agouti. In particular it involves female orchid bees, which pollinate the Brazil-nut flower. Their male partners never go near the tree. Instead they fly to orchids in the forest, where they scrape off the plant's waxy, scented secretions and fly about en masse to attract the females with their perfume cloud. Thus, without orchids there would be no orchid bees, male or female; and without the female orchid bees for pollination, there would be no Brazil-nut trees, either.

This, says Ortiz, explains the mystery of why plantations of Brazil-nut trees never thrive. While the trees will grow outside the forest, the orchids will not; and the trees need the orchids because they need the bees. In this way, one of the giants of the forest is dependent on some of the smallest occupants of the forest floor. For this reason too, Brazil-nut trees will not grow anywhere outside the Amazon: they need their little helpers and, without them, they are nothing. Brazil, as Charley's Aunt had it in the Victorian farce of the same name, 'is where the nuts come from'. And that is not just a memorable dramatic line: Brazil is the only place the nuts come from. Worth some 70 million dollars a ye[illegible] vacuum-wrapped exports from South America, Brazil nuts are a rare example of a fruit of the forest that remains loyal to and profitable for its original habitat.

But the tree's interaction with the forest is far from done. Its cannon-ball pods crash on to the forest floor during the wettest season of the year, at a time when rain penetrates the canopy above and reaches the ground. After the agoutis have done their sharp-toothed business, the empty pod husks are left behind on the forest floor. There, they create their own micro-ecosystems, like a flooded forest in miniature. Unique species of mosquitoes and damselflies found nowhere else use the small pools inside the pods to lay their eggs. So does a species of poison-arrow frog, which relies almost exclusively on rain-filled Brazil-nut pods for water to get through the tadpole stage and emerge as a fully grown frog.

So: no agoutis, no Brazil-nut trees. No orchids, no orchid bees, and again no Brazil-nut trees. And no Brazil nuts means no frogs. Without these arrangements, says Ortiz, 'this forest would be very different. The Brazil nut is the dominant tree here. To a great extent, they are responsible for what this forest looks like.'

* * *

There is no end to the bizarre interactions between species in the forests. A dung beetle in Borneo specializes in rolling away balls of dung from orang-utans and laying its eggs in them. But another beetle smells out those infested dung s in them, and its grubs then kill those e. And that's not all: an orchid has devel- f partly decomposed orang-utan dung to ond beetle, which turns out to be the only le of pollinating that orchid.

An stories go on. A species of fly injects its eggs into the bodies of leaf-cutter ants as they carry their cargo. But, in reaction, other tiny ants hitch a ride on the backs of the first ants to defend them from the flies. Meanwhile there are parasitic mites that sit on flowers waiting for the chance to run up the beaks of visiting humming birds.

Making these kinds of discoveries demands obsessives – scientists who are prepared to believe the often barely believable evidence of their own eyes, whether it is tiny agoutis able to crack the toughest nuts in the forest, or moths with tongues longer than elephants' pollinating hard-to-get orchids. So, in this gallery of bizarre science for a bizarre world, step forward Kim Bostwick of Cornell University in New York, who used ultra-fast cameras, taking hundreds of frames a second, to photograph male manakin birds. She did this in an effort to discover how the birds make a weird cracking sound, like a firecracker, that attracts females. They do it, the frames reveal, by drawing their tail feathers up behind their head and cracking them like a whip, so fast that no slower camera can see what is going on. At up to 110 beats a second, this is twice the speed of a hummingbird's wing-beats. Only a rattlesnake can move its muscles faster. In fact, the manakin has evolved unique bone and muscle structure in order to perform this ritual courtship display.

Right up there with the obsessives is Bostwick's colleague Leonida Fusani, whose videoing of a golden-collared manakin found that what looked like a simple act of jumping to the ground and bounding back up actually comprised a high-speed somersault, followed by the clapping of wing-tips behind its head before the bird landed and took off again.

And take a bow, too, Hans Bänziger. A Swiss-German who grew up in Italy, Bänziger has spent much of his life in the rainforests of Thailand. There he discovered moths with a taste for tears. At first his findings were dismissed as nonsense – or the result of a single observation of a deranged moth. Not so. He kept looking and now it turns out that, unbeknown to generations of moth-lovers, hundreds of moth species across the tropical forests drink the tears of water buffalo and deer, horses and tapirs, pigs, elephants, and sometimes even sup the lachrymose emissions of people – something to which Bänziger himself can attest, after an itchy night in the Thai jungle a few years ago. 'The sensation of having one's eyes sucked is somewhat unpleasant, rather like having a grain of sand in the eyeball,' he says.

Moths suck tears at some risk to life and limb: even a batting eyelid could be enough to kill them. But they seem to choose the tears of the more docile animals, generally leaving aggressive carnivores alone. As well as in Thailand, they do it elsewhere in Asia, in the African jungle and in parts of South America. In New Guinea, they drink the tears of pigs that were introduced to the island only a few centuries ago. Did the moths come with the pigs, or did the local moths develop a new taste? What do they get out of it? It probably isn't water, since they drink tears even in the rain. According to Bänziger, it is more likely to be salt, and perhaps proteins. It turns out that tears are not these moths' only unusual drink. They also like saliva, nasal mucus, sweat, urine and the anal blood of mosquitoes. And what do the animals that yield up their bodily fluids get out of it? Darwin would have said that they must get something – but that is a nut that today's researchers have yet to crack.

HIDDEN BEASTS

How much do we know about forest elephants of Africa? A lot, you would think. We are talking, after all, about the biggest, best-known animal in the jungle. But most of what we know is actually about the forest elephants on the plains of East and southern Africa. The forest elephants of Central Africa are more mysterious. Amazingly, it was only in 2001 that scientists concluded that the forest elephants, which make up about a third of the continent's total elephant population, were in fact a different species from their cousins on the plains. The African forest elephant is smaller, with more rounded ears and smaller tusks, which are straight and point down, so they do not get tangled up in the dense vegetation. In some ways they are more like the Asian elephants. Both have four nails on their hind feet and five nails on their front

WHERE THE NUTS COME FROM

Natives in the Amazon, like the Kayapo, have planted Brazil-nut trees for millennia. They both eat them raw and brew the bark of the tree to make a tea that has been found beneficial for countering liver disease. But the nuts did not catch on in Europe, even after the German explorer Alexander von Humboldt discovered their distinctive taste on his journey through the Amazon in the 1790s. The outside world really learned to love them only in the wake of the rubber boom of the mid-nineteenth century, when exports of the two Amazon products increased in unison.

Nut-collecting worked well in the winter wet season, when the latex stopped flowing. For a while the collectors did both jobs, and the nuts found a ready market in Europe and North America. When the bottom fell out of the Brazilian rubber business, the collectors stayed behind in the forest to collect the nuts. The nut has come a long way since then – from a favourite nibble in British drawing-rooms of the nineteenth century to a flavouring in Ben and Jerry's ice cream, oil for cooking and an ingredient in hair conditioners. The husks are also burned to produce mosquito-repelling smoke and carved into trinkets. But the half a million or so nut collectors, known as *castaneros*, have come rather less far. They are still fleeced by the agents of the handful of exporters operating out of the port of Belem, just as they were in the late nineteenth century.

feet, in contrast with the African savannah elephant, which has a four- and three-nail configuration. This is something a French zoologist noticed back in the 1930s, but it somehow got ignored for more than half a century.

Genetically, the two species of African elephant are more distant from each other than a tiger is from a lion or a horse from a zebra. And, far from being recent cousins, they diverged genetically more than 2 million years ago. As *National Geographic* asked when the discovery was made: how do you miss a whole elephant species?

And we are talking about much more than genetics: in behavioural terms, too, the African forest elephant is a 'different animal'. It is much more secretive and elusive than its cousin, and surprisingly light-footed when making its way through the forest. It has a different diet, browsing endlessly on fruit taken from the trees. It often appears to be a loner, rarely given to collecting in large herds, which would, in any case, be difficult in the jungle. Yet that would be to misjudge them, for they are, in truth, highly social animals. For instance, they form great herds that migrate through the jungle at night. Steve Blake of the Wildlife Conservation Society was the first to discover these great treks and recorded them using radio-collars strapped to the elephants and starlight cameras. The discovery has hugely important implications for developing conservation strategies to protect these animals.

The secret sociability of the African forest elephant is a key to another remarkable and only recently discovered characteristic. With apologies to Mohammed Ali, we might call it their 'rumble in the jungle'. They communicate with each other over long distances using a low-frequency vocal throb that is transmitted through the ground like seismic waves. The throb is too low to hear, but can be picked up by the elephants' feet. It can travel for miles through the forest, much further than vocal sounds passing through the air, which are muffled by thick forest vegetation. It has been suggested that elephants use this calling system to attract mates, keep track of calves and to guide each other towards good foraging grounds.

According to biologist Katy Payne of Cornell University, who is developing ways of using this long-distance communication system in order to count elephants in the forest, while most humans would not be aware of a forest elephant until it is almost upon them, another forest elephant would know about it from a great distance. Payne has set up microphones in the forest to hear elephants that cannot be seen, and installed software to distinguish one call from another; she says it beats tramping through the jungle counting piles of dung, a method of elephant-counting recently discredited because the formula used to estimate the 'decay rate' of elephant droppings turns out to be unreliable.

But the mystery of the world's forest elephants goes further than the African jungles, for another elephant subspecies has recently been unearthed – this time from the jungles of Malaysian Borneo. Detailed genetic analysis of elephant dung found on the forest floor has revealed that the 'pygmy' elephants that live in the deep rainforests there are a different subspecies from the main Asian elephants that we know from India and elsewhere. They are smaller and more mild-mannered, with bigger ears, longer tails and straighter tusks than their cousins, either in the neighbouring island of Sumatra or in mainland Asia.

It was once thought that these Borneo elephants were the ancestors of elephants shipped to Borneo from the Asian mainland by the British East India Company about three hundred years ago. They certainly had an order from a Sultan who ruled what is now the Malaysian province of Sabah in the far north of the island, and where most Borneo elephants are found today. But genetic analysis shows that the thousand or so elephants on the island today have been on their own there for far longer – 300,000 years, according to estimates made at New York's Columbia University. They went there of their own accord and eventually got cut off during one of the periodic rises in sea level that created the islands of southeast Asia. The shipment for the Sultan seems to have died out.

* * *

Many other large mammals of the jungle also lead surprisingly secretive lives. Take tapirs: these great, shy, nocturnal beasts can grow up to 3 metres long, are shaped a bit like a pig but weigh as much as a horse. They are well studied in Latin America, but the species that lives in southeast Asia, the Malayan tapir, has barely been studied at all. One of the few people to take an interest is Jeremy Holden of Fauna and Flora International, who has been setting camera traps for them in the forest. These night travellers trigger infra-red beams that set off both a camera and a flash, and take their own photograph.

Wild cats also prove surprisingly difficult to capture on film. The Borneo bay cat, for example, is only about the size of a domestic cat and skulks around the rainforests of Borneo. In 2003, press reports said it had been 'rediscovered' after being thought extinct. Well, not quite. It had not been recorded as extinct, but nobody had seen it alive for a while. The last one had been found, mortally wounded in a snare, two years previously. Then on 27 June 2003, a Malaysian biologist captured an image in a photo trap and, four months later, a Russian biologist did the same. These were the first pictures taken of the cat in the wild. The Borneo bay cat was back.

Equally mysterious among the cats of the Asian jungles are the Sumatran tigers, of whom nobody took much notice for a long time. Hunters and explorers revered tigers in general, of course, calling them the 'king of the jungle'; but when, in the late nineteenth century, they drew up their lists of different sorts of tigers, they saw the Sumatran specimen as pretty indistinguishable from other Asian tigers. They spent a lot of time distinguishing between different mainland subspecies, like the Siberian, south China and Bengal tigers, and along the way grudgingly accorded the Sumatran tiger similar subspecies status, and left it at that. It seemed a shame when the Sumatran tiger began to be seen less and less often in and around the Sumatran rainforests, but, even as the forests began to disappear fast in the 1980s, nobody thought much about the beleaguered Sumatran tiger.

Then Joel Cracraft, of the American Museum of Natural History, stepped out of the jungle one day with a new genetic analysis. Far from being a mere subspecies, he said, Sumatran tigers should be regarded as their own distinct species. Not only that, he declared, most of the mainland cats are, genetically speaking, pretty much identical. Genetically, there is little or nothing between a Siberian and a Bengal tiger. Cracraft argued that, in reality, there are only two tiger species: mainland and Sumatran. The debate continues and some tiger experts want to stick with the old classification. They argue that, however genetically distinct the Sumatran tigers may be, they can breed with mainland tigers. Whatever happens, however, the status of the Sumatran tiger is in the process of being transformed from lowly migrant to king of Sumatra. 'The Sumatran tigers were isolated on Sumatra after a rise in sea level created the island approximately six thousand years ago,' says Cracraft. They have spent enough time breeding on their own to be genetically distinct.

So, step forward the Sumatran tiger. But just how many are there? Nobody knows. The official estimate is that there are four hundred left. Some say there could be thousands still holed up in the diminishing jungle; and new research has found Sumatran tigers living happily enough among the forest logging grounds and oil-palm plantations that generally take over when the forests are chopped down. Others think there could be many fewer, as farmers, loggers and miners invade the island's last tracts of forest, nibbling away even at its small protected reserves, and poachers go after a prey that can fetch three thousand dollars for a pelt and 150 dollars for a penis. With most Sumatran tigers in zoos thought to be interbred with other species, the only true, pure-bred Sumatran tigers may turn out to be in the wild. Only in 2004 were the first successful moving images of a Sumatran tiger captured.

* * *

Despite such hasty genetic redefinitions, scientists had thought that the days of finding new big animals in the jungles were gone. But, despite the fact that hardly a scrap of the planet is unoccupied by humans and everywhere is under the constant eye of passing satellites, the past decade or so has seen a renaissance in the discovery of large mammals. It began in 1992, when John McKinnon of WWF and his Vietnamese partner Do Tuoc took advantage of the first peace in the country for half a century to explore the remote forests of the Annamite mountains between Vietnam and Laos. There, local hunters showed them skulls with strange straight horns from an animal they called the saola. When they tracked the animal down, it turned out to be a cinnamon-coloured ox. It was not just a new species, but a new genus – an entirely new branch of the mammalian tree. Two years later they found a new species of deer, called a giant muntjac, which looked disconcertingly like Walt Disney's Bambi. A couple of smaller species of muntjacs, new to science, also turned up.

But then the locals got wise to the intense interest among foreigners in their local fauna, with distressing results. They came to McKinnon with some weird, twisted horns, rather like high-rise motorcycle handlebars, which they claimed to have found in a local market. The horns were instantly, and all too publicly, declared to be the remains of some unknown species of mountain goat. Suspicions grew when no further specimens were found and, in 2001, other researchers revealed that the horns were genetically identical to those of local cattle and had been heated, bent and carved to their outlandish shape.

Nevertheless, the new species keep coming. Not to be outdone by McKinnon's work for WWF, the organization's arch-rivals in the US, Conservation International and the Wildlife Conservation Society, went looking too. And they have been finding. In 1999, Alan Rabinowitz of the WCS turned up the world's smallest deer in remote jungle in northern Burma. Just 50 centimetres high, it is called by local hunters a 'leaf deer' because it is so light they can carry its carcass through the jungle on a particular type of leaf. A month later, Rob Timmins, a British forest biologist at the WCS, revealed that he had found three striped rabbits in a food market in upcountry Laos. Till then, the world had known about only one striped rabbit – from the jungles of Sumatra. And that had been seen alive only once in eighty years.

Then, on an expedition for Conservation International to the Vilcabamba Mountains on the edge of the Peruvian Amazon, Louise Emmons of the Smithsonian Institution disturbed a weasel dragging some prey off the path she was following. The weasel ran off, but the prey turned out to be a tree rat the size of a domestic cat – apparently a relative of the tomb rats that Incas kept as pets and took with them to their graves. 'Another metre either way in that dense vegetation and I'd have missed it,' admits Emmons.

But, as ever, some of the weirdest stuff comes from the Amazon, where Dutch zoologist Marc van Roosmalen says he can't stop finding new species. A veteran of twenty years of research in the Amazon, van Roosmalen began his odyssey when native traders visited his home in the jungle capital of Manaus. Knowing his enthusiasm for weird animals they brought him a tiny dwarf marmoset in a tin can. 'I saw this little thing screaming like mad. I knew it was new,' he says. So he went looking for its home, which turned out to be a previously unresearched stretch of swampy rainforest surrounded by giant caymans that ate anything that tried to leave. It was, he says, like Arthur Conan Doyle's *Lost World*. The animals there were as effectively cut off as on any remote island. Van Roosmalen believes that, besides several other new marmosets, he is on the track of more unknown monkeys and new species of porcupines, tapirs and maybe even a new black jaguar.

'Many people are convinced that our knowledge of the world's large creatures and cultures is all but complete. They are wrong,' Rabinowitz wrote in *Beyond The Last Village* – a book on his Burmese days. 'Our world remains a wondrous place. You can still find your way to the end of the last dirt roads, where maps show nothing but blue and forest green, where there are still worlds to be explored and mystery is waiting.'

ALIENS FROM THE JUNGLE

As humans invade the rainforests, they are bringing out denizens of the jungle that run riot in the outside world. Many quickly die, of course, when removed from their natural environment; but some thrive and, with no natural enemies to keep them in check, became major pests.

Take the venomous brown tree snake. A native of the jungles of Queensland, New Guinea and especially the Solomon Islands, it is adept at curling up in a small, inconspicuous ball. Sometime in the 1940s, a few of these creatures seem to have used this skill to hitch a ride with some Japanese soldiers on a plane to the Pacific island of Guam. There, the snake, which can grow to 3 metres long, found plenty of food and no predators. When the Americans took over the island base after the Second World War, they found that it was infested with the alien snakes. Repulsing the Japanese proved easier than getting rid of the snakes, who are still there in greater numbers than ever. One recent snake census found thirteen thousand of them in one square mile. They have eaten their way through most of the island's forest bird population, occasionally kill babies in their cots and have triggered hundreds of power cuts by slithering along cables and causing short-circuits.

Islands are susceptible to aliens because they generally have few native species that can fight back, and are sitting targets for anything that can get a foothold. One profligate island invader is the yellow crazy ant from the forests of West Africa. It has been on the march for a while, invading African islands such as Zanzibar and the Seychelles before hopping to Hawaii in the Pacific. But its latest target is the Australia-run Christmas Island in the Indian Ocean.

Nobody knows how it got there, but it has taken to the place, setting up residence on the island's few surviving tracts of rainforest and forming a series of 'super-colonies', from where the ants go foraging for food, with alarming results. They are currently decimating the island's world-famous population of red land crabs: by one estimate, the ants killed and ate 3 million crabs in just eighteen months. The crabs are essential to the island's forest ecosystems, recycling fallen leaves to create forest soil. Now, with the last crab likely to be eaten soon, biologists fear that the ants already have their next meal lined up – the island's even more famous population of the Abbott's booby, a seabird that nests nowhere else on Earth.

A third invader from the forests is a weed, the water hyacinth. Victorian rubber barons were the first to notice this big-leafed water weed with its large purple and violet flowers growing prettily across the backwaters of Amazon tributaries. It reportedly adorned a garden pond in one of the rubber barons' mansions in the Peruvian jungle city of Iquitos sometime in the 1880s. From there, someone – perhaps the notorious Julio Cesar Arana, who divided his time between Iquitos and Biarritz in the south-west of France – shipped some weed out and, before long, water hyacinth was the green foliage of choice in ornamental garden ponds across the imperial globe.

Inevitably, the weed escaped into the wild and proved a doughty invader. It typically doubles its biomass in twelve days and can now be found infesting waterways, irrigation channels, reservoirs and lakes across fifty countries on five continents. There is barely a river or lake in Africa that has escaped infestation. It grows faster than any other plant, destroying fish-spawning grounds and deoxygenating the water. On Lake Victoria in East Africa, it clogs ports and water inlets and forms thick, impenetrable mats, several miles across, that harbour crocodiles, poisonous snakes, mosquitoes and the snails that carry bilharzia.

In the Amazon, the weed's spread is held in check by native insects and microbes. So now scientists have gone back to Iquitos on a new biological hunt. Where once they sought rubber, now they are looking for predators to kill water hyacinth. They have found that some flies lay larvae that eat the stems of the hyacinth. Others carry plant pathogens that will attack the weed. If one of those insects can make its home in African water-hyacinth mats and get chomping, then they are in business.

WHAT ELSE COULD BE OUT THERE?

There are many stories of half-seen creatures in the jungle. Many, no doubt, are myths and hoaxes – but perhaps not all. Enough scientists have taken an interest to give a name to the discipline of pursuing them: cryptozoology. Apparently serious researchers occasionally head into the swampy Congo forests, east of Ndoki, in search of Mokele Mbembe, which local Bayaka tribes say is a large, aggressive animal with a pointed horn that lives deep in the swamps. Some outsiders have dubbed it, somewhat improbably, the Congo Dinosaur. More likely it is a rhino. Normally found in the savannah plains, this would be a rare, frightening and memorable sight for any rainforest inhabitant.

Researchers continue to wonder if the Brazilian forest still contains one last three-ton giant ground sloth. And what of the orang-pendek, supposed inhabitant of the Sumatran rainforest? Enthusiasts are convinced that this yeti-like creature exists and say it is part of the mythology of local tribes.

According to David Chivers, deputy chairman of the UK-based Fauna and Flora International, 'We are talking about hundreds of these creatures. The most likely scenario is that it is a new species of ape: a gibbon or an orang-utan.' Evidence includes hair samples that do not seem to be from orang-utans, chimpanzees, gorillas or humans, and the cast of its footprint that seems to be rather like (and perhaps is) the heel of a human and the ball of an orang-utan.

THE HIGH FRONTIER

The idea of jungles being a massive tangle of impenetrable vegetation is largely myth. Undergrowth is found mainly at the borders, along river banks and clearings and wherever light can penetrate. Because humans are responsible for most of the clearings these days, the irony is that what we think of as archetypal jungle is, in fact, often man-made. Inside dense jungle it is too dark for much to grow. As the biologist Edward Wilson puts it, the interiors of rainforests are 'green cathedrals', in which 'there is almost never a need to slash a path with a machete through tangled vegetation' and it 'is so dark a flashlight is needed to study it closely'. For the real action you need to look up, into the branches. The most biodiverse and densely inhabited area is not on the ground but in the canopy, exposed to the full impact of the sun and rain.

Welcome to the forest canopy. It is a world that was virtually unknown until a few years ago, and remains largely unsurveyed. And yet it is probably the most important wildlife habitat on the planet. This 'high frontier' is, by some measures, home to 40 per cent of all the plant species. Perhaps half of all life on Earth could live up there, largely unseen and unknown. It was only when researchers like the Smithsonian's Terry Erwin began counting bugs in the canopy that our estimates of global species numbers rose from five to thirty and even to 100 million. About a quarter of all our insects are reckoned to live only up there. Just as the rainforests are the hub of the planet's biodiversity, so the hub of the rainforest is in the canopy. And just as the first European invaders of the rainforests five hundred years ago marvelled at their discovery, so the privileged few who have penetrated the canopy can claim a similarly unique perspective on the living planet.

Scientists have always had their suspicions about the rainforest canopy. In 1917, American naturalist William Beebe declared that 'another continent of life remains to be discovered, not upon the Earth, but one to two hundred feet above it, extending over thousands of square miles . . . There awaits a rich harvest for the naturalist who overcomes the obstacles – gravitation, ants, thorns, rotten trunks – and mounts to the summits of the jungle trees'. Around that time, Edwardian explorers wearing pith helmets took early forays by firing ropes into the high branches with cannons and suspending vast quantities of block and tackle to raise themselves in bosun's chairs as high as they dared to go.

Now that world is being explored more intensively and with a little more facility. In the 1980s, a new generation of dare-devil scientists – more acquainted with extreme sports than butterfly nets and binoculars – began to explore the canopy. These early pioneers used crossbows to get up high, shooting cables and ropes over branches and heading upwards like mountain climbers, then, once aloft, they threw ropes from tree to tree. Their aim was to join the party, swinging from tree to tree like the lemurs and monkeys, squirrels and snakes, who had made their homes in the branches.

Heroes still live up here: perhaps not Tarzan, but improbable legends nonetheless. Roman Dial is one. Back home in Alaska, where he also goes mountaineering and white-water kayaking, Dial is one of the top skiers, bikers and river-runners. But when he is finished with sports on the ground he heads for the Borneo rainforest canopy, where his pioneering methods of rigging and climbing have made him one of the world's foremost canopy 'trekkers'. During a typical day's work in Borneo, he goes up and down 350 metres of ropes, penetrating to different layers of the canopy, mapping the forest in 3D and, along the way, collecting thousands of insects by setting up fogging machines that spray an insecticide into the canopy.

But perhaps the glory days are over almost before we know about them. Today, the slingshots and crampons of the canopy trekkers are gradually being replaced by more expensive kit, so that the less intrepid scientists can make it into the branches, too. A global network of canopy scientists is persuading governments to fork out millions of pounds for planting giant construction cranes in the forest. They are floating beneath balloons and even hiring airships that can navigate their gondolas straight into the canopy. This is the new biological frontier: a new great age of jungle exploration is just beginning.

So far, ten cranes have been erected in a handful of the world's most important forests, including Barro Colorado island, the headwaters of the Orinoco in Venezuela and the Malaysian Borneo province of Sabah. They are like satellites landing on a distant planet, providing a tiny snapshot of another world. A high-tech canopy tent known as Ikos allows researchers to work, eat and sleep in the canopy for up to a week in Gabon, Madagascar and Panama. And for those who don't fancy making the journey themselves, there is the Flybot, a highly manoeuvrable and silent remote-controlled camera that can be operated from the ground to fly through the canopy capturing images of a world never seen before.

And what a world. On the forest floor you can see only a few yards in the gloom, but up in the canopy you are bathed in light and can see for miles. According to Nalini Nadkarni, president of the International Canopy Network, 'Our findings have already fundamentally altered the notion of what a forest is.' Here, the branches of trees are covered in moss, on which plants and even young trees grow in what amounts to giant window-boxes hundreds of feet up in the canopy, where clouds hang in the air and everything is drenched in moisture. In this suspended ecosystem there are earthworms, snakes, beetles, ants and plants that never see the earth of the forest floor, and snakes and beetles, ants and plants, fed on by pollinating birds.

The canopy is home to myriad creatures, from jaguars, sloths and monkeys to tree frogs and eagles. The branches form highways along which canopy animals find food and mates. As they used to say of empty spaces on medieval maps, here too be monsters. This, for instance, is the domain of the biggest, most rapacious living thing in the rainforest – the strangler fig. This vegetable monster begins life simply enough when a fruit-eating bat drops its seed onto a tree branch, where it germinates and begins to feed off moss. Bathed in the sunlight and watered by the copious rain falling into the canopy, it grows fast, at first sending out tendrils that explore the canopy for moisture and nutrients. But this is not enough: as it grows bigger, it gives up on its canopy hosts and lets tendrils down; these creep round the trunk of the tree till they reach the forest floor, where they form roots.

Having done that, and having no further need for its host, the fig strangles it. Boosted by nutrients from the forest floor, the tendrils straggling round the trunk thicken and tighten, so the tree can no longer grow. Meanwhile up in the canopy, the strangler fig eventually towers above the tree that gave it life, depriving it of light and starving it to death. Even giant Brazil-nut trees cannot escape its deadly embrace. But this is murder with a purpose for the forest: the hollow trunk, which is eventually all that remains of the tree, becomes a perfect home for numerous animals and insects. In a mature forest, the giant stranglers are the largest organisms left, overseeing all, consuming all. They are the most dramatic evidence that, in the jungle, life begins at the top.

RAINMAKING

The canopy is not just a vast weird zoo, or even just the epicentre of the world's biodiversity. It is where the rainforests – the planet's largest body of living matter – breathe. Think of great rainforests like the Amazon and the Congo jungle as single vast organisms sitting astride the tropics, inhaling and exhaling through the canopy. They are the biggest living entities on the planet, with a very strong breath and they make our atmosphere what it is. What the rainforests breathe out, we breathe in. They are the planet's atmospheric filter, its air conditioner and its sprinkler system. If they get bad breath, or if the rainforests disappear, we are in big trouble.

The rainforests' biggest planetary job is as the great engines of one of the most basic and important biological processes – photosynthesis. This process happens inside every plant cell: it is where sunlight is converted into plant matter, and it is how plants grow. In the process, photosynthesis also has a fundamental impact on the atmosphere that we breathe: vegetation is composed largely of carbon, and photosynthesis absorbs huge amounts of carbon dioxide from the atmosphere. This ability of vegetation – and of rainforests in particular – to absorb carbon dioxide through the pores of leaves helps to stabilize our climate and acts like a planetary thermostat. Here's how.

Carbon dioxide is a greenhouse gas. Every molecule of carbon dioxide in the atmosphere helps trap heat close to the Earth's surface, and this helps to keep the atmosphere warm. But too much carbon dioxide will cause us to overheat, and that has become a growing concern as humans burn more and more fossil fuels, like coal and oil, which are made of the carbon remains of ancient rain-forests. Burning fossil fuels pours this ancient carbon dioxide back into the atmosphere, and the world is already warming as a result. But it would be warming much more without the current rainforests, which absorb about a third of the carbon dioxide put into the atmosphere in this way, so helping to keep us cool. Climate scientists believe that if you took away all the rainforests, the world's temperatures would soar by several degrees as all the carbon dioxide they contain ended up in the atmosphere.

Researching the chemical make-up of the 'forests' breath' and what it means for the rest of life on Earth is a major research activity today. It is one of the biggest tasks for the equipment aboard the new cranes towering over forest canopies in all the major rainforest regions. These sniffers of the forests' breath have a lot to test for. Forests inhale and exhale other things besides carbon dioxide. They breathe out oxygen, of course, helping maintain atmospheric levels just high enough for us to breathe easily, but not so high that the entire world spontaneously bursts into flame. They also help produce hydroxyl (a natural chemical that cleans other pollutants from the air), carbon monoxide, methane and ozone – all of which are critical to the chemistry of our atmosphere.

It seems that the composition of our atmosphere would be very different without the rainforests – perhaps literally unbreathable.

The rainforests are also the planet's rainmakers, constantly recycling the water that falls from tropical clouds. As much as two-thirds of the rain falling into the canopy of tropical rainforests never reaches the ground. Instead it either evaporates from the leaves right back into the air, or is absorbed, used and then released again by the vegetation – a process known as transpiration. Chop down the rainforests and the rain will instead fall to the earth and run away to the sea. There will be no recycling of the water back into the air. So, downwind, the air will become drier, more rainforests will die and deserts will start to grow. So the rainforest canopy also makes the weather that sustains the forests. Without the rain there would be no forest, but without the forest there would be much less rain.

Rainforests are so big and so dominant that they don't just adapt to their tropical environment: they make that environment. As Andrew Mitchell, director of the Global Canopy Programme, puts it, the rainforest 'fills our lungs with oxygen, breathes rain across the globe, and sifts pollutants from the air. It is a complex guardian, tending the health of the planet, as delicate and important to the Earth as the lining of our lungs is to us.' Take it away and the environment changes for all of us.

PRIMEVAL GARDENS

The primeval, virgin forest is a potent modern myth. But the more researchers look, the less virginal it seems. The remains of the Mayan pyramids of Central America and the Angkor temples in Cambodia are just the best-known examples of a global phenomenon of organized settlements and dazzling urban architecture in the jungle. Hidden in the bush, it turns out, are the remains of ancient structures as large as the Great Wall of China and ancient suburbs as big and as sprawling as any in southern California. The key to these civilizations was an ability to 'garden' the rainforest that nobody could attempt today. The skills were lost as suddenly as the great cities died and the jungle reclaimed its own.

SUNGBO'S EREDO

Sometimes the biggest things can be the hardest to notice. How else do you explain the failure of generations of Nigerian and European archaeologists to spot Sungbo's Eredo, the ramparts of a thousand-year-old kingdom not an hour's drive from Lagos, a city of 10 million people? The ramparts are very big: the earth wall and ditch are 160 kilometres long and tower, in places, as high as a seven-storey house. They come complete with guardhouses, moats and garrison barracks, and enclose an area the size of Greater London. But it took a British geographer from the University of Bournemouth to recognize them.

Patrick Darling first saw Sungbo's Eredo when he spotted a strange earth mound while driving his car along the Pan-African Highway in southern Nigeria. He stopped the car and set off on foot into the undergrowth to explore. What he found that day was the jewel in the African civil-engineering crown. Sungbo's Eredo was constructed around one thousand years ago and is the outstanding example among thousands of ancient city ramparts, boundary embankments and ditches that stretch through former jungle across Nigeria and much of West Africa. Darling says that tropical landscapes are littered with ancient earthworks that dwarf more famous ancient mega-structures such as the pyramids of Egypt or even the Great Wall of China. He has made it his life's work to unlock the secrets of these Cinderellas of the archaeological world.

Darling did not, strictly speaking, discover Sungbo's Eredo – local African villagers and pilgrims knew about it all along. Eredo means a ditch, and locals believe that the earthworks were built a long time ago for Bilikisu Sungbo, a fabulously wealthy but childless widow who wanted a monument to her rule of the area. Thousands of Islamic pilgrims come each year to pay homage to Sungbo at her shrine, which lies just inside the ramparts. Local legend has it that Sungbo was in fact the Queen of Sheba, who appears in both the Bible and the Koran as the wife of Solomon and the ruler of a wealthy kingdom in Arabia or Ethiopia. That is probably a myth. Sheba is supposed to have lived around three thousand years ago, so there is a two-thousand-year discrepancy in dates. Probably the Nigerians who first wrote up the local oral histories linked everything to biblical stories or the Koran when they could, and the idea of a black African queen led them straight to Sheba.

The reality may not be so romantic, but it is still impressive. Sungbo's Eredo is Africa's largest single ancient monument. The ramparts encircle an area 40 by 35 kilometres, with sheer walls rising to 10 metres in places. Darling has unearthed the charcoal remains of fires set to clear the forest before the walls were built. Radio-carbon dating reveals that the charcoal is around 1,200 years old, give or take a century or so. 'This pushes back the date of any known African rainforest kingdom by hundreds of years,' he says. 'It is the first definitive proof that there was substantial state formation in the African rainforest at roughly the same time as in the savannah.'

So who masterminded the construction of Sungbo's Eredo? Tracking events more than a millennium ago in a country with no written tradition is not easy. The ramparts seem to mark the boundary of an ancient kingdom, probably populated by the Ijebu, one of several groups of people collectively called the Yoruba, who still live in Nigeria. Sungbo's Eredo could have been built by the Awujale dynasty, which still reigns in Ijebu-Ode – a mud-walled Yoruba town in the heart of the Eredo. The dynasty's oral history encompasses fifty-seven kings and queens. Before 1760 the dating is hazy, but the dynasty probably began around AD 1150. So it looks as though the Eredo was already under construction when the Awujale kings took over – and the mystery remains.

It would have taken millions of hours of toil to build Sungbo's Eredo; but Darling is most impressed by the skill displayed in surveying and construction. Whoever built it had the 'ability to retain a coherent masterplan in thick rainforest, frequently interrupted by great fingers of swamp,' he says. 'Without compass or aerial photographs, they were able to keep the ramparts on course whatever the obstacle.' Today, much of the bare earth dug and heaped early in the ninth century – around the time that Charlemagne ruled in Europe – is covered in green moss. The ditches reek of rotting vegetation and echo to the sound of croaking frogs. The whole thing is engulfed by trees, undergrowth and fetid swamp. It lies overgrown and apparently forgotten, rather like an old railway embankment in the English countryside.

There are hints that early explorers may have known about this huge structure. In 1505, the Portuguese sea captain and explorer Pacheco Pereira – who is credited by some with discovering Brazil seven years earlier – reported the existence in the Nigerian jungle of a 'very large city called Geebuu, surrounded by a great moat'. The conventional view is that this was Ijebu-Ode, where the Portuguese bought slaves in return for brass bracelets. But Darling says that it could have been a reference to Sungbo's Eredo itself. Then the written records go silent. Generations of slave traders and cocoa-plantation managers and loggers passed through the area without seeing a thing – until a brief report in 1959 by British anthropologist Peter Lloyd in *Odu*, an obscure journal on Yoruba studies. 'I found Sungbo's Eredo when researching customary land laws,' recalls Lloyd. 'When I looked at aerial photographs, it was quite visible, partly because it was still covered in trees, unlike much of the surrounding land.' Lloyd's paper went unnoticed by archaeologists, however, and only resurfaced when Darling went exploring.

For Darling, Sungbo's Eredo is just the most startling element in an even bigger story that began thirty years ago when, as a young teacher in Nigeria, he was coaching a state cycling team. While out training one day, he noticed that his team were cycling past vast networks of Nigerian earthworks around Benin City, which is 250 kilometres east of Lagos and a long way from Sungbo's Eredo. Nobody appeared to know what these mounds and ditches were, but local officials offered him a small scholarship to research them. Since then, whenever local politics and his teaching career have allowed, Darling has visited Nigeria to carry on with the survey, using little more than his bicycle and an orienteering compass.

No single structure in the Benin network is as large as Sungbo's Eredo, but together they dwarf it. They extend for some 16,000 kilometres in all, in a mosaic of more than five hundred interconnected settlement boundaries, and cover an area of 6,500 square kilometres. In all, they are four times longer than the Great Wall of China, and consumed a hundred times more material than the Great Pyramid of Cheops. They took an estimated 150 million hours of digging to construct and are perhaps the largest single archaeological phenomenon on the planet. No wonder Darling has still mapped only around a tenth of them.

Darling's radio-carbon dating suggests that the Benin earthworks were constructed as a rolling programme, most of which was carried out between AD 800 and 1500, roughly the same time as Sungbo's Eredo. At the core of the complex, the enclosures are small, surrounding individual clan communities. In these areas, there may be 5 kilometres of earthworks in every square kilometre. But, towards the periphery, they are large, with a circumference of up to 100 kilometres, and a wide no-man's-land between the enclosures. Darling believes the development of the Benin earthworks probably marks the gradual migration of farmers as they extended their domain deeper into the jungle, enclosing ever-larger stretches of land.

With the aid of anthropologists such as Dan Ben-Amos, a specialist in African folklore at Pennsylvania State University, Darling has started to piece together the social as well as the physical structure of the settlements. They include places such as Ohovbe, a migration camp 7 kilometres east of Benin City; a rash of settlements, all called Oka, that share a common earthwork boundary; and the city of Udo, which once rivalled Benin and boasts 10 kilometres of ramparts containing 500,000 cubic metres of earth. The studies are helping to unearth the earliest history of this area. In Ishan Forest northwest of Benin, slash-and-burn farmers first cleared the rainforests at least two thousand years ago. Nearby, the Edo people dug the earliest banked compounds in the Benin complex some 1,500 years ago at a time when local tribes were switching to more settled methods of farming. Perhaps, Darling speculates, this change gave them more time for digging out the walls of the settlement, or maybe both developments arose after they started making iron implements.

The question remains: why go to all that trouble to enclose the land with giant moats and embankments, whether at Benin or Sungbo's Eredo? Why build the walls so high and erect guard towers and what looks like military barracks? The earthworks must have had some role in defence, says Darling. But a deep, dark rainforest is a very different place from an open plain, where most of the world's defensive ramparts are built. In the rainforest, visibility is only a few metres and warfare is about poisoned arrows and sneaking about, not cavalries and pitched battles. Some have argued that the ramparts were built to keep out a rather different enemy – elephants. 'It is hard for us to appreciate what it was like for farming communities trying to survive before the invention of modern weapons, when surrounded by hundreds of elephants,' says Juliet Clutton-Brock, an archaeo-zoologist recently retired from the Natural History Museum in London. 'Elephants must have been terrifying, and worth any amount of effort in building ditches and walls to keep them out.'

Perhaps just as important as keeping enemies and large animals out, the earthworks would have kept the people of these kingdoms together and under control. 'All over this region, you find that the societies that had ramparts seem to have been more cohesive and to have survived for longer against outsiders,' says Darling. The ramparts may also have had an important spiritual role, especially for people who came from outside the rainforest and who may well still have been afraid of it. According to local traditions, the earthworks represent the boundary between the real world inside the enclosure and the spirit world outside. The inhabitants buried charm pots beneath gateways in the earthworks, and applied potions to the earth walls to ward off evil spirits. To this day, there is an annual ceremony to appease the spirits of the gateways; and people throw gifts to the spirits in the ditches as they return home from funerals held outside the walls. The ditches, in particular, are a kind of limbo land. 'In Britain we have the phrase Devil's Dyke, and it's rather like that. Evil spirits live there,' Darling says.

As intriguing as the question of what these structures represent is the question of why they have been ignored for so long. 'Darling's is a major piece of work, of importance for the whole of Africa, and it hasn't had the attention it deserves,' says John Alexander, an archaeologist at the University of Cambridge. 'Unfortunately, nobody else, either from Nigeria or abroad, is investigating these earthworks.' Foreign archaeologists, Alexander says, seem to prefer the cooler, drier climates of Egypt, Ethiopia and southern Africa. 'For every tonne of carefully sifted soil in Egypt, less than a teaspoon has been glanced at in West Africa.' And because virtually none of the world's great archaeologists have been to the region, they assume there is nothing to find there. Even Nigerians have been strangely reluctant to venture into the field, says Lloyd. 'I suppose earthworks weren't very exciting. There is not much chance of finding grand tombs of jewellery or anything.' There are three universities with archaeology departments within an hour of Sungbo's Eredo, he says. But, in forty years, nobody has gone to investigate.

Darling's work is helping to piece together a radically new history for the rainforest regions of West Africa, and it chimes in with research going on all round the tropics. The idea of virgin rainforest, largely untouched by native people, is giving way to the realization that humans have sculpted even deep jungle for perhaps thousands of years. The theory that civilization got going only in drier climates – and that rainforest dwellers were happy to live a semi-nomadic life of hunting, gathering and sporadic slash-and-burn farming – is being kicked into some very long grass.

HUMAN REMAINS

Some of the pioneers in this radical reappraisal of the rainforests come from the fringes of academia. They are more like the adventurers of old. So, swinging down from the trees is Mike Fay, an American veteran of Central Africa, who has survived a plane crash, confrontations with armed rebels and being gored by a large elephant. He has campaigned for, set up and run one of the continent's largest forest national parks, the Nouabale-Ndoki park in Congo-Brazzaville, and persuaded Gabon to establish thirteen more. And he is also one of the few people in modern times to walk right across the swamp rainforests of the Congo basin in Central Africa to the Atlantic coast. The 2,000-kilometre trek took him fourteen months. It sounds like one of the great Congo journeys by Cameron and Stanley a century and a quarter ago, but Fay did it in half the time it took Stanley. And, while Stanley took more than 350 staff and 7 tonnes of arms and equipment, carried by a half-mile-long column of porters that could only be halted by blowing a bugle, Fay headed off with a couple of local guides and not much more than a small knapsack containing a spare pair of sandals.

Fay has been an ardent environmental activist but, at the end of his odyssey, he conceded that his jungle quest had not, strictly speaking, been through virgin forest at all. 'I have never come across a virgin forest in Africa. It is obvious that man has been a player for a very long time,' he says. Wherever he walked, he said, the soil was littered with oil-palm nuts. These are not native fruits of the forests, but are probably the remains of plantations established by human farmers some two thousand years ago. The jungle where he found the nuts covers some 10,000 square kilometres in the northern Congo basin. Most of it today is uninhabited. Yet back then, at a time when the Romans were creating their empire across Europe, it was peppered with clearings containing oil-palm plantations.

The Congo jungle, evidently, is not all it seems. When Fay flew his Cessna plane over the forest, he kept noticing small grassy mounds in the clearings that he is convinced are the remains of ancient farming systems. When he waded the swamps along the Sangha river in northern Congo, he turned up vast numbers of pottery fragments. Perhaps most dramatic of all, in the dense forest of the Central African Republic, he found the remains of iron smelting and strong evidence that, a thousand years ago, the locals were chopping down trees on a large scale to fuel the smelters. His findings have been corroborated by French archaeologist Richard Oslisly. In the Ogooue valley in the middle of modern-day Gabon, Oslisly has uncovered hilltop iron towns that were also in the smelting business more than two thousand years ago, but which disappeared, abruptly and mysteriously, about 1,500 years ago. This was not just a farming landscape in the forest: it was an industrial landscape in which 'extermination of the forest likely occurred over vast areas'.

What happened to end these industrial cultures? Fay believes they probably ran out of trees to fuel their furnaces, but not everyone is convinced. Others argue that there was a population crash right across the Congo at that time, which requires some other explanation. Perhaps there was a dramatic change of climate or a major epidemic – an early version of AIDS or the Ebola virus, maybe. In any event, the jungle closed in over some calamitous event. It seems, too, that the modern human population of the Congo is just a shadow of its former self, and that, despite present-day fears about deforestation, the Congo rainforest is probably currently close to its historically maximum extent.

Our ideas of primeval African forest don't hold water. Kathy Willis of the University of Oxford asked in the title of a recent paper: 'How virgin is virgin rainforest?' She answered for Central Africa that 'much of this region underwent extensive habitation, clearance and cultivation beginning around three thousand years ago and ending 1,600 years ago in the aftermath of a population crash'. It seems clear, agrees conservation scientist John Oates, of the City University of New York, that 'human disturbance has been one of the dominant factors affecting [African] forest structure and composition in recent millennia'.

Some believe there has been a series of human population crashes within the Central African rainforest. Some are clearly tied to the influence of Europeans – most notably the height of the slave trade in the eighteenth century and the 'rubber terror' of Belgium's King Leopold. And there is intriguing evidence that something also happened around three hundred years ago. Barend van Gemerden of Wageningen University in the Netherlands examined the ecology of apparently pristine rainforest in southern Cameroon. When he looked closely it did not seem to be pristine at all. Most of the older trees in the forest, those more than three hundred years old, were of species that today grow only in the large clearings created by shifting cultivators. By contrast, most of the younger trees were of species that preferred closed forest with few clearings.

The odd age distribution, he decided, could have come about only if the forests were once much more densely populated than they are today, with farmers cutting and burning the forest on a large scale. Something must have changed in southern Cameroon about three centuries ago to cut that population and reduce the interference in the forest. This would have been around the time when Europeans were cruising the shores of the continent to barter for slaves and gold, before they had penetrated into the interior. But it was before the height of the slave trade. Did the early slave trade trigger greater events within the interior than previously known? Or were there already great social disruptions that may have delivered a harvest of slaves to these coastal traders?

LOST WORLDS

If the African rainforests are far from virgin, what of the other jungle regions of the world? Much of the Asian rainforest remains unstudied by archaeologists, but one exception is the area around Angkor Wat in central Cambodia. The Angkor civilization arose around AD 800 and grew, reaching its height in the twelfth and thirteenth centuries, when it ruled a huge area of southeast Asia stretching from the South China Sea to Thailand and the uplands of Laos to the Malay peninsula. But the civilization abruptly crashed following the sacking of Angkor by the armies of the Thai kingdom of Ayutthaya in 1431. Since then, the extraordinary Angkor temple complex has been discovered and largely forgotten several times. It was first found by Portuguese friars looking for people to convert to Christianity in the 1580s. 'There are ruins of an ancient city in the jungle of north Cambodia, inhabited by ferocious animals, which some say was built by Alexander the Great or the Romans,' wrote Marcello de Ribadeneyra. The French rediscovered it in the 1860s and again in the 1920s. It disappeared from international view again during the long Indochina Wars, but today it is once again a great tourist attraction.

But the civilization was not just a series of stunning temples in the jungle. In the past five years, studies of satellite images from space have revealed for the first time that the temples were simply the ceremonial heartland of a vast and densely populated suburban area that has been described as rather like modern-day southern California. It was, according to detailed research by a team from the University of Sydney, 'by far the most extensive pre-industrial city in the world', covering at least 1,000 square kilometres, and twenty times the size of Manhattan. The satellite pictures began with radar mapping from the *Endeavour* space shuttle in 1994. They showed the full scale of the networks of roads, canals, reservoirs and rice paddy fields that spread from the shore of the great central Cambodian lake of Tonle Sap into the far distant Kulen Hills. Those images are now being followed up by detailed archaeological research on the ground, which is revealing settlements containing miners and weavers, boat-builders, blacksmiths and a large salt industry. The entire urban area in the jungle must have housed hundreds of thousands of people, many living in houses along the canals and on raised embankments that carried roads to the edges of the city.

What is less clear is where this civilization came from – and why it crashed. Was it a lone oasis of urban society amid empty jungle or, as seems more likely, simply the pinnacle of achievement of a series of societies in the forests of southeast Asia? Bio-archaeologist Lisa Kealhofer of Santa Clara University in California has found evidence of widespread farming in the forests of Thailand eight thousand years ago, during what she calls a 'lost bronze age'. Maybe such societies were the predecessors of the great urban civilization of Angkor, she says. Meanwhile, modern researchers believe that the Angkor urban area may finally have collapsed when river diversions that filled its rivers failed, possibly after being silted up. Perhaps there was some kind of ecological crisis triggered by deforestation. But all this remains speculation.

The great temples, and the evidence of a large urban zone around them, show the power of humans even a thousand years ago to transform a jungle into a urban landscape. But travel round the temples, as hundreds of thousands of tourists now do each year, and you see, too, the power of the jungle to claim its own. In temples such as Ta Prohm, vast pyramidal constructions of stone have been turned over by invading trees, their huge buttresses pushing over walls several metres thick. Meanwhile, the roots of strangler figs grow out of the top of walls and press down to split and surround the stone as surely as they would have done a tree in the forest.

* * *

Cut to Latin America, where the most famous jungle remains come from the ancient Mayan civilization. Great pyramids loom out of the jungle as a highly visible reminder of a culture that levelled large areas of the forests of Guatemala, Honduras and southern Mexico, starting some three thousand years ago, and converted it to an urban landscape that lasted for two millennia. Besides pyramids, this long-lasting civilization built vast temples, established seats of learning that made major advances in mathematics and astronomy, pioneered the cultivation of maize and built giant reservoirs, viaducts and canals for irrigation. It was, in its day, one of the most densely populated regions on the planet, but it collapsed during the ninth century AD and has since been largely reclaimed by the jungle.

When American adventurers John Lloyd Stevens and Frederick Catherwood stumbled on the Mayan acropolis at Copan in Honduras in 1839, they found a 'lost city' eight miles long and lined with temples and pyramids. 'The city was desolate. No remnant of the race that built it hung in the ruins. It all lay before us like a shattered bark in the midst of the ocean . . . her crew perished and none to tell from whence she came. All was mystery, dark impenetrable mystery . . . I knew that a new chapter in history was unfolding before us, that we were entering upon new ground and that a great mystery was there to be solved,' wrote Stevens. So he bought it, paying the locals fifty dollars.

Other parts of the great empire have been discovered since. In the 1980s, René Muñoz and Charles Golden went searching in the Guatemalan jungle and found the remains of half-discovered buildings not previously mapped – plus a 1930s tractor, apparently left there by researchers more interested in bagging Mayan loot than recording the 'lost city'. In the jungle, and even in the modern era, whole civilizations can get lost as well as found.

But what became of the Mayan civilization, which abruptly disappeared about 1,200 years ago? Some blame drought for the Mayans' demise; others blame overpopulation or the delayed ecological impact of deforestation. But more interesting, perhaps, is that it survived so well and for so long. And that it was not alone. Once seen as some kind of mysterious aberration, it now seems as if the Mayans, like the empire builders of Angkor in far-away Cambodia at around the same time, were far from alone in the jungle.

Husband and wife David and Deborah Clark were for many years directors of the remote La Selva forest research station in the Central American state of Costa Rica. Home to tapirs, sloths, jaguars, howler monkeys and much more, the station is one of the four reference research centres for presumed pristine tropical rainforests recognized by the US National Academy of Sciences. Scientists have been working there for years to describe its unchanging ecosystems and burgeoning biodiversity. Well, that is what they thought they were doing until a few years back, when the Clarks emerged from the lush forest to report their latest discoveries. They had not found new species of beetles or orchids, or even a tree bark that yielded a cure for cancer. They had found instead, buried in the soils in the heart of the forest, charcoal, corn pollen and farm tools.

The whole area, they reported, had been used for slash-and-burn agriculture for most of the past two thousand years. The high forest of La Selva, apparently the epitome of pristine forest, turned out to be no virgin. Far from being ancient jungle, most of it was recent regrowth. The place was little more than a garden gone wild after being abandoned by farmers perhaps less than five hundred years ago, probably when Europeans first invaded Central America in the wake of Columbus. 'There is no such thing as virginity out there,' Deborah Clark told the *New York Times*.

REVISITING THE AMAZON

And then there is the amazing story of the Amazon. Surely here, in the greatest rainforest on the planet, virginity can be discovered – nature red in tooth and claw? That's what everyone thought until an American oil prospector called Kenneth Lee first climbed aboard a beach buggy and bounced across the lowlands of Baures in the Bolivian Amazon in the early 1980s. This landscape is today covered in rich grasslands that flood for long periods each year, with thousands of forest islands. After a while, Lee began to wonder why he was bouncing so much in the grass. On closer inspection, the landscape appeared to be corrugated. It was composed of a remarkably symmetrical series of ridges and trenches stretching as far as the eye could see. From time to time, he came across higher ridges that looked like roads, and wider depressions that seemed as if they might once have been canals. He began to think that this must be a man-made landscape and would take occasional visitors to see his 'lost civilization of Baures', which he believed must extend into the rainforest proper to the north and east.

And so it has proved. Subsequent research by Clark Erickson of the University of Pennsylvania has found tens of thousands of kilometres of raised banks across the Bolivian Amazon that he believes were dug by humans. By corrugating the flooded fields, farmers created ridges on which they could plant their crops, clear of the floodwaters and also of highland frosts, while also collecting water for irrigation in the dry

In the nineteenth century, King Leopold of Belgium claimed the whole of the Congo (including this stretch in the Kahuzi-Biega national park) as his personal kingdom – with fearful results.

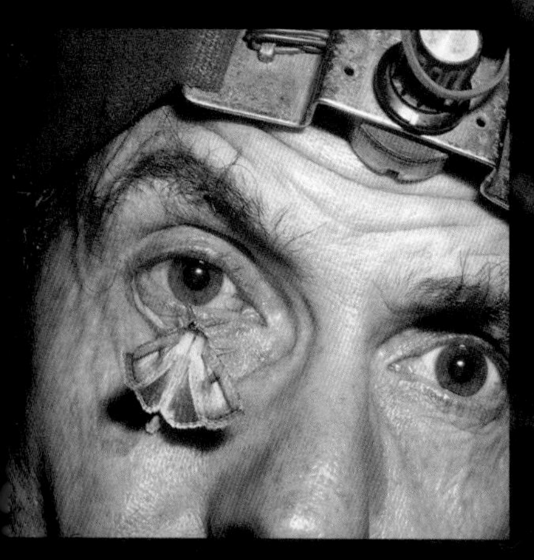

Top | **During the search for El Dorado, gold pendants like this one from ancient Mexico were melted down and shipped home to Europe.**

The jaguars of the Pantanal are
the largest in the world.

Middle | A tear-drinking moth goes to work on its discoverer, Hans Bänziger.
Bottom | Nameless wonder: a new species of Teagueia orchid, as yet unnamed,
was uncovered by Lou Jost of Ecuador's Pastaza Valley in the late twentieth century.

Above | **Brazil and its inhabitants, as mapped around 1520 by Portuguese cartographer Pedro Reinel.**

Above | **Shaving the planet – Clear-felling on hill slopes is the world's most destructive form of forestry.**

Main | **Asia's only great ape – orang-utan numbers have halved in the last decade as loggers have moved in.**

Main | **Angkor Wat – 'the most extensive pre-industrial city in the world' – is now being reclaimed by the jungle.**

Top | Edible pods from petai trees sustain the Talang Mamak people in Sumatra.
Bottom | Madagascar's rosy periwinkle held the chemical secret to fighting childhood leukaemia.

The enormous silverback Mlima – which means 'mountain' in Swahili – is head of the Dzanga-Sangha focus group.

season. It was a flat-land equivalent of the ancient practice of terracing hillsides. The digging and earth-moving involved in creating these structures, he says, is 'comparable to building the pyramids. They completely altered the landscape.'

Today, that landscape is all but empty, but the evidence of human existence remains both in the land and in the artefacts they left behind. Each of the forest islands contains ample evidence of human habitation, such as pottery, human bones and charcoal. The mounds cover several hectares and rise some 5 metres above the surrounding land in a series of wide terraces. Some would have been capable of housing thousands of people. They, too, must be man-made, Erickson says. Meanwhile, the larger causeways show up clearly on satellite images, running straight across the plains. Some have called them the markings left by visitors from outer space. But Erickson says they are roads, 3 or 4 metres wide and about a metre above the rest of the land – enough to provide a dry walking surface.

Erickson has found something else. 'When we began taking our Cessna plane over the area, we noticed strange low earth walls zigzagging across the savannah and between the surviving islands of forests. They were separate from the corrugating of the fields, but did not make sense either as roads.' Then Erickson noticed that the low walls had small funnel-like openings whenever they changed direction. 'I realized that these matched the descriptions of fish weirs in the historical literature about the Amazonian peoples. These are fences made of brush to trap migrating fish.' What he had stumbled on was a vast system – estimated to cover

500 square kilometres – of fish ponds and weirs. When the grassland flooded in the rainy season, the people captured the fish, put them in ponds and kept them for eating during the dry season. They were fish-farming on the edge of the rainforest.

Erickson believes that the fish ponds and raised fields stretching across the plains for thousands of square kilometres could have sustained maybe a million people. Buried charcoal in the roads and mounds suggests that they were created up to two thousand years ago. Could it be true? Surprisingly, there is evidence from early Spanish accounts that seem to describe just such settlements. An expedition to Baures in 1617 described entering towns along causeways that could take four horse riders abreast. Jesuit records confirm this and suggest that some islands and causeways remained in use into the eighteenth century, before being left to regrowing forest and populations of tapirs, peccaries and deer. 'Some people want to preserve the forests,' says Erickson. 'That is fine by me, but there is no way they are pristine. Every feature of this land is man-made.'

* * *

After Lee and Erickson upset the Amazon myths in Bolivia, along came Anna Roosevelt, an archaeologist at the Field Museum of Natural History in Chicago. In 1990, she was digging on Marajo, a flat, forested island, twice the size of Wales, in the mouth of the river Amazon, which was recently the setting for the Brazilian version of the reality-TV show

Survivor. Roosevelt expected to find evidence of a few village communities on this strategic but inhospitable island. Instead she found hundreds of large earthworks on the forest floor, each some 20 metres high and covering up to a square kilometre. Remains inside the mounds showed that they had been erected between AD 400 and 1400. The mounds had been the focal points for numerous urban centres. She concluded that, a thousand years ago, the entire island was crossed by roads and irrigation and drainage networks and dotted with large towns, in which maybe 100,000 people lived and worked. Roosevelt called it 'one of the outstanding indigenous cultural achievements of the New World'.

Today, Marajo is just on the fringes of the rainforest region so, since Roosevelt's discoveries, other researchers have looked for evidence of similar civilizations further west in the heart of the rainforest. Michael Heckenberger of the University of Florida at Gainesville went to one of the deepest, darkest areas of continuous tropical rainforest, in the Upper Xingu region of the state of Mato Grosso. It is an area that has been inhabited by the Xinguano people for at least a thousand years – something we know because of the bits of ceramics they have left behind on the forest floor. But what he found was that the ground has not always been forest floor. In fact most of this primeval forest had been cleared at least once, and perhaps several times, by the Xinguano for farming.

This was not an ostentatious urban society with pyramids and so on, as in the Mayan civilization. Instead, it seems that what is today one of the largest tracts of rainforest in the world was, till relatively recently, a chunk of tropical suburbia. 'This really blew us away,' says Heckenberger. 'It's fantastic stuff. Everyone loves the "lost civilization in the Amazon story". But what the Upper Xingu shows us is that Amazon people organized in an alternative way to urbanization. We shouldn't be expecting to find lost cities. But that doesn't mean they were primitive tribes, either.'

Heckenberger focused on one of these suburban areas called Kuikuro. Here he found the remains of nineteen settlements. They were about 4 kilometres apart, each on a raised area a couple of kilometres long, rather like Roosevelt's mounds on Marajo, and linked to the others by a system of wide boulevards up to 50 metres across. At the heart of each settlement was a big plaza, with roads radiating from it towards a surrounding moat, like a huge castle in the forest. In the surrounding sometimes swampy land between the settlements, there were bridges and dams, dykes and causeways, canals and ponds, manioc gardens and surviving forest patches housing orchards and places where medicinal plants and other 'fruits of the forest' grew.

It was, Heckenberger says, 'a highly elaborate built environment, rivalling that of many contemporary complex societies'. The settlements were permanent, and all seem to have been thriving communities until the Europeans turned up at the end of the fifteenth century. After that, the settlements were abandoned and the high forest returned to reclaim the suburbs. Some areas, says Heckenberger, are still regrowing; others are almost back to what they must once have been. The strange truth is that by inadvertently wiping out the Indian populations, it was the Europeans who created the modern Amazon rainforest. In both the Amazon and the Congo rainforests, local civilizations seem to have crashed shortly after the first substantial contacts with western Europeans.

HOLOCAUST

Why did we not know about these societies before? Did the conquistadors not notice them as they passed through the Amazon? The strange thing is that some did notice. Beside the ravings about El Dorado and tribes of warrior women, several chroniclers noted briefly – almost in passing and in a matter-of-fact way – how they encountered settlements containing tens of thousands of people along the banks of the Amazon, headed by warrior kings and ruling large areas.

Take this from Francisco de Orellana's famous first expedition down the Amazon in 1542, in search of El Dorado. As he entered the Rio Negro, one of the Amazon's biggest tributaries, he wrote in his journal: 'There was one town that stretched for 15 miles without any space from house to house, which was a marvellous thing to behold. There were many roads here that entered into the interior of the land, very fine highways. Inland from the river to a distance of six miles more or less, there could be seen some very large cities that glistened in white and besides this, the land is as fertile and as normal in appearance as our Spain.'

This city was never seen again. When conquistadors returned to the area a few decades later, they found only a few inhabited Indian settlements. Within a century, all trace of these societies had vanished and explorers found only small bands of Indians deep in the jungle. Later, such stories of great cities in the jungle were derided. A new orthodoxy held that civilization was purely a preserve of cooler lands, whether at temperate latitudes or, when it occurred in the tropics, confined to highland regions like the Andes. But the truth, says Erickson, is rather different. 'The Spanish came to the Amazon looking for El Dorado. They didn't find it, but I think it was there all along.' There were no piles of gold in the jungle, but there were great cities. In the end, it was not the conquistadors who deluded themselves, but those scientists who followed and thought they knew better.

Knowledge of these civilizations largely died out as the people who made them retreated back into the bush to escape from the Europeans, or were destroyed by Old World diseases that hitched a ride with the human invaders. Probably millions died in the decades after the Spaniards came and, as disease literally decimated their societies, the survivors fled into the jungle. It seems likely that while the farmers and ranchers and metal-makers and priests and scholars who must have made up these societies gave it all up to become hunters and gatherers, the European invaders and despoilers were barely aware of what was going on in front of their eyes. For American Indians, the arrival of Europeans had a similar impact to what Europeans imagine might happen in the event of a nuclear holocaust. The primitive nature of some tribes still being 'discovered' from time to time in the Amazon rainforests is due in large part to the arrival of Europeans. Likewise much of the 'virgin' forest of the basin may be regrowth following the exterminations of the conquerors. If anyone made the jungle and their 'stone-age inhabitants', it was the Europeans.

Fragments of the old world live on in folk tales in the forest and, most intriguingly, in language. For instance, the Guaja tribe, who still live in Maranhao in the eastern Amazon, say they once grew maize; and they have words for other crops like sweet potato and yam. Botanist William Balee of Tulane University in New Orleans has investigated their story. They seem to have survived better than many communities before succumbing. 'Between the late 1700s and 1860,' he says, 'something happened. It could have been an epidemic from the Old World – and it drove some groups to extinction and others were so reduced that they could not maintain their agriculture. Eventually they lost their domesticated plants.'

For Erickson, this debunks some potent myths. The greatest of these is 'the myth of the pristine environment', in which the landscapes of the Americas were largely undisturbed by nature until the arrival of Europeans. Biologists have assumed, almost as an article of faith, that the areas with the greatest variety of species must be the most natural. Far from it, he says. 'In fact, we find that high biodiversity is clearly related to past human activities, such as creating clearings, burning and gardening.' Balee agrees. Biodiversity in areas of the Amazon farmed by pre-Colombian societies is, he says, 'probably higher today because of human occupation, use and management of the land than if people had not been there'. The wonderfully productive forest that biologists have lauded as an example of wild fecundity – filled with Brazil nuts, lianas, palms, bamboo and other economically valuable forest species – may be the deliberate invention of man rather than the accidental beneficence of nature. 'We now know,' says Erickson, 'that much of what has traditionally been recognized as wilderness in the Amazon is the indirect result of massive depopulation after Europeans brought diseases, slavery and war.'

GARDENING IN THE RAINFOREST

How did scientists get things so wrong for so long? In large part, archaeologists have not found remains in rainforests because they haven't gone looking: the plains of Kenya, after all, are rather more easily investigated than the jungles of the Congo. And they have rather assumed that because they found the jungle environment hostile, others before them would have avoided it, too. Equally, many of the detailed observations about rainforest regions in the interiors of the continents – the ones that established our ways of looking at them – took place during a period when, especially in the Americas, native populations had been decimated by disease and conquest. As Stephen Pyne from Arizona State University puts it: 'Virgin forest was not encountered in the sixteenth and seventeenth centuries; it was invented in the late eighteenth and early nineteenth centuries.'

The forests then looked to many as if they had always been empty, even though many had only recently been emptied. And yet, even at that time, the most acute observers were aware of something different, that the forest inhabitants were not so much noble savages as the traumatized survivors of past societies. Alexander von Humboldt wrote of a 'degenerate race' in the jungle, which he went on to describe as 'the feeble remains of nations which after being long scattered in the forests, have been again immersed in barbarism'.

One further reason why modern scientists have been reluctant to believe that great civilizations lived in the rainforests, especially in the Amazon, comes from their studies of rainforest soils. Betty Meggers of the Smithsonian Institution in 1971 wrote a highly influential book on Amazonia subtitled 'Counterfeit Paradise', in which she argued that the jungle soil was too poor to sustain large populations and that they could never have fed themselves. This chimed in with the views of many botanists such as Paul Richards – the British author of a classical text, *The Tropical Rainforest* – who, in the 1950s, said that native people had had 'no more influence on the vegetation than any of the other animal inhabitants'.

So why are Meggers and Richards wrong? One of the most persuasive pieces of evidence about the extent of human occupation and transformation of the Amazon rainforest is the widespread presence in inhabited areas of the forest of what archaeologists have dubbed 'black soil'. This is superficially very like the regular 'yellow soil' that sceptics like

Meggers said could not sustain complex societies. But mixed with it is a mulch of organic waste and partly burnt plant material, rather like charcoal. Both fertilize the soil. This mulch has permeated the black soils with micro-organisms that ensure that it regenerates itself even as it is used for cultivation. It is also often full of pottery shards and other detritus of civilization, much of it now dated to two thousand or more years ago. This has convinced many researchers that the 'black soil' is not just a convenient natural soil: it is in all probability man-made.

Bruno Glaser of the University of Bayreuth in Germany says, 'the really important point here is that the soils contain charred residues. That is different from the residues of [natural] burning. Both of them improve soil fertility, but burning residues don't last for long, while charring residues have a long-term effect on soil fertility, acting over centuries. It's at least as good as manure. In some places we know that Indians successfully farmed land containing black soil for 2,500 years or more.' Indeed, scientists are becoming so convinced that this stuff is man-made that they have begun mapping the black soil as a surrogate for the extent of human occupation and transformation of the jungle.

If all this is true, then the ancients had a far better idea of how to farm in the jungle than their modern counterparts. Burning is what modern slash-and-burn farmers do. It leaves behind fine ash that washes away in the rain. Most of it is gone within four years. But charring, which leaves organic material permanently in the soil, seems to be a skill that has been lost. To this day, the black soils are more fertile and provide good crops of manioc, maize and bananas. Local farmers still prize it, because yields are so much greater than when they plant on 'yellow soil'. In places, locals dig up the black soil in the forest and transplant it to improve their garden soils, like Europeans buying bags of peat at the local garden centre.

Black soils pervade the banks of the Amazon and its key tributaries, like the Rio Negro and the Tapajos. The story away from the rivers is less clear, but some researchers believe it will turn out to extend deep into the heart of the Amazon and to occupy perhaps a tenth of all the rainforest soils, an area the size of France. But, whatever the extent of black soil turns out to be, it is clear that the Amazon is, in places, capable of sustaining large populations; and that, across the Amazon, farmers have – probably for millennia – improved soils using fire, planted crops like manioc, palm and Brazil nuts, and cleared forests to let in the sun and grow corn. And that, in all probability, the Amazon was dotted with urban centres and criss-crossed by networks of causeways and irrigation canals at the same time as the Greek empire flourished in Europe.

* * *

The truth is that farmers do not merely destroy forests; they often embellish and improve them, according to American anthropologist Darrell Posey, who spent a decade living with the Kayapo Indians in the Gorotire Indian Reserve in the heart of the Brazilian Amazon. Posey, who died in 2001, pointed out forcefully that, within the forests, the ancestors of his hosts had planted many 'forest islands' containing fruit trees. He discovered that almost everything that scientists learnt about the rainforests, the locals knew before, and often in much greater detail. The average Kayapo child, he said, can name and identify sixty different sorts of bee. The Kayapo classify bushland into fifteen distinct types, and have purposes as food, medicines and so on for more than six hundred species of plants. Around their cultivated clearings, they plant particular types of bananas that are enjoyed by wasps that would otherwise feed on the crops. And before they abandon the fields, they plant groves of Brazil nuts and fruit trees, which they return to harvest for decades afterwards. They re-forest grasslands and transplant medicinal plants and fruit trees to plots near their villages, along trails and river banks and within hunting grounds. Some trees are planted specifically to entice animals that they want to kill.

Posey pointed out that, like the great European botanical prospectors of the nineteenth century, the Kayapo have travelled huge areas of Amazon forest in search of useful plants. Nor does it seem that the Kayapo are exceptional: it is simply that Posey took the trouble to document their rainforest 'gardening'. There are, in all probability, large tracts of forest in the Amazon that are the deliberate creation of humans. Probably other rainforests are the same.

Ancient rainforest dwellers were brilliant gardeners. Until recently, it was thought that New World farming began in the cool highlands. But all that changed with the discovery

MYSTERIES OF THE SOUTH PACIFIC

Sometimes the past is much closer than we think. New research carried out in the Solomon Islands, in the South Pacific, by Tim Bayliss-Smith of the University of Cambridge has found that at least one island, New Georgia, had a population two centuries ago that was twice what it is today. There were towns with up to seven thousand people, where now there are only small villages amid the trees. 'This apparently pristine island rainforest may be a consequence of regeneration over the past 150 years following population decline and migration to coastal regions,' says Kathy Willis.

And sometimes the past is much smarter. The tribes of New Guinea are widely seen today as among the most primitive in the world. They are caricatured as a 'Stone Age' people: head-hunters living in the remote mountain forests of one of the world's largest and least populated islands, and sometimes entirely uncontacted by the outside world. So far as we yet know, the people of this island have never built cities; but it now seems that the ancestors of today's 'Stone Age' people were once at the cutting edge of the then new technology of cultivation. Some seven thousand years ago, the people of New Guinea were clearing lowland rainforest and replacing it with intensive banana plantations. It is very likely that they were the first people to work out how to cultivate bananas.

that farmers in the forest swamps of coastal Ecuador were nurturing prize vegetables as long as ten thousand years ago. Archaeologist Dolores Piperno, of the Smithsonian Institution in Panama, discovered in rubbish dumps of that date microscopic crystals from the rinds of cucurbita plants – which include marrows, gourds and cucumbers. The crystals, Piperno reported, were far larger than those produced by wild versions of cucurbita, and much more like modern cultivated varieties. She concluded that the eccentric English tradition of cultivating super-marrows must go back far longer than anyone thought – to the jungles of Ecuador in the waning days of the last ice age.

DEMON FARMERS AND MYTHS OF DEFORESTATION

Just as we have myths about forests, so, too, we have myths about their disappearance. While nobody will deny that deforestation has been widespread throughout the tropics in recent years (and we will describe that in a later chapter), some of the numbers are distinctly woolly – and the consequences of getting them wrong are more than statistical. Take the case of Madagascar.

William McConnell of Indiana University looked at official estimates of forest cover on this great rainforest island off the

African coast, which has some of the greatest and most valued biodiversity on the planet. To get a handle on current trends, he looked at two studies done in the 1890s. They put the forest cover at either 20 million or 12 million hectares. Quite a big difference. You would think we would probably have got better at counting trees in the intervening century, especially given the importance we attach to the unique species of Madagascan forests. Well, no. McConnell found two studies from the 1990s: one said Madagascar had 5.8 million hectares of forest left; the other put the figure at 13.3 million – proportionally an even bigger difference. And what does this say about trends? Take one pair from each decade and you would reckon that three-quarters of the forest had gone. Take another pair and you could reckon that forest cover has increased.

Was Madagascar once completely covered in forest? Lots of conservationists imply so. It is the 'natural' vegetation there, they say. But recent studies of lake sediments on the island show profuse remains of grass pollen, especially in the central highlands. So when the US government's Agency for International Development reports that 'slash-and-burn agriculture has destroyed over 80 per cent of the tropical forest cover', can we believe it? And what are we to make of it when the Wildlife Conservation Society, based at Bronx Zoo in New York, pitches for cash donations for conserving Madagascan forests by advertising that the remaining rainforests are a 'Paradise in peril' because slash-and-burn farming has 'transformed large areas of forest into wastelands'?

If we cannot be sure of these claims, is it right for international conservation groups and aid agencies to advise the government there to ban slash-and-burn farming and create protected forest areas to keep the farmers out? At this point, conservation and power politics make an unedifying brew. As McConnell puts it: 'A misdiagnosis of the nature and causes of environmental change in Madagascar can only undermine conservation efforts and cause unnecessary hardship for its residents', who, as we are constantly reminded, are among the world's poorest.

* * *

Or take the forests of West Africa. Melissa Leach of the University of Sussex argues that environmental claims that the West African rainforests have been disappearing fast in the past forty years may be wide of the mark. 'Often there is more forest cover today than a generation ago,' she says. In past times, outsiders made broad-brush estimates of the size of undisturbed forest without noticing the large areas within the forests that were inhabited and farmed by locals. Today, the opposite mistake is made, with surveyors ignoring the many forest islands dotted through the farming landscape.

There are, Leach agrees, real problems with large-scale industrialized logging of forests and clearing for commercial farms and mining. Global markets mean that these activities are undoubtedly happening on a scale never seen before in the forest. But local farmers act differently – and perhaps more similarly to their ancestors at times when the African countryside may have been as heavily populated as it is today. Many of today's forest islands are planted and tended by villagers for their own purposes. Rural communities, says Leach, are often custodians rather than destroyers of forests.

Her studies found that, in the West African state of Guinea, for instance, locals cultivate and maintain forest islands around their villages, planting trees as physical protection against the winds, as cover for shade-loving crops like coffee, for fruits and nuts and medicinal plants, to help protect water supplies, as reserve fodder for their animals, and often as sacred groves with a clear religious as well as practical purpose. Sometimes these islands are the remains of old and larger forests; but sometimes they are deliberately created.

Leach quotes the official figures for the West African state of Sierra Leone. Early in the twentieth century, British foresters made a detailed study of the country and concluded that it had only 150,000 hectares of forest left. On the other hand, Norman Myers, a noted modern rainforest conservation scientist, quotes a report from 1945 that the country had 5 million hectares of 'little disturbed forest' – thirty times more. Which figure should we use to compare with the current situation? Myers uses his to demonstrate that deforestation is rampant. But use the other figure and the country appears more forested at the end of the twentieth century than at the start.

The impact of such misunderstandings about the history of our surviving rainforests can be pernicious, especially for the people who today live in and around those forests. For the myth of the virgin forest has given rise to the myth of the demon farmer. Traditional slash-and-burn farmers become routinely denigrated as destroyers of the forest – and not just in the propaganda of Western environmentalists: they are pilloried by their own governments and thrown out of national parks and forest reserves that are arbitrarily declared across their land.

And yet, look a bit more closely and you find that many farmers are assiduous protectors of forest patches. One ancient tradition that survives to this day is the protection of sacred groves, which are found all over the tropics. In West Africa, the villages of Ghana have a long tradition of maintaining sacred groves as protected areas of forest. The groves still have real religious significance for them: traditionally, it is bad luck or worse to clear the groves or hurt animals within them. As a result, the groves have become sanctuaries for wildlife such as monkeys and antelopes, while supplying medicinal plants and wild foods and protecting water supplies and even local climate. By one recent count, there are still more than a thousand sacred groves in Ghana. Some are threatened by the extension of farming lands as village populations grow, but now, in an unusual merging of traditional and state law, villagers in central Ghana are trying to revive them.

The villages of Baubeng and Fiema in central Ghana are on the northern fringe of the West African rainforest region. Between them sits a sacred grove. Farmers there know that their crops grow best on land close to the forest, because the forests ensure more water in the dry season and harbour insects that keep down weeds. Unfortunately, that encourages individual farmers to push their fields right up to the edge of the forest – or further. In recent years, disputes have erupted in the villages when farmers have cleared part of a sacred grove.

The grove is threatened with death by a thousand cuts.

But a village teacher in Baubeng, Daniel Kwaku Akowuah, devised a plan to save the grove by combining traditional religious teaching with modern conservation ideas. He proposed turning the grove into an animal reserve that would attract tourists. The spirituality of the grove for villagers is strongly tied up with the presence there of 350 mona and black and white colobus monkeys, so Daniel persuaded the local district assembly to reinforce traditional village authority by creating the Buabeng-Fiema monkey sanctuary at the centre of the 4.5-square-kilometre sacred grove. New bylaws gave villagers the power to arrest anybody who cut down trees in the grove or killed the monkeys.

The laws created tensions at first. There were arrests and court convictions, and one outside timber company tried to bribe Daniel to allow tree felling. But most villagers have supported the scheme, especially now that tourists are coming. Today, the sanctuary has a guest house and a listing in the West African *Rough Guide*, a popular guide book that is sold all over the world. The status of the monkeys as a good omen for the village has been reinforced, with their economic value complementing their spiritual significance. And the nearby village of Dotobaa has been so impressed that it is reviving a sacred spring in a piece of relic forest where monkeys and other animals once lived. One farmer who regularly clears rotting leaves from around the spring, Yaw Gjan, believes his work could bring back animals – even lions – one day. 'When we get the forest, there will be wonders here again.'

* * *

A revised version of the rainforest myth holds that while the great rainforests may indeed have sustained great urban cultures, those cultures were ecologically dangerous and ultimately doomed. By clearing the forest, their inhabitants interfered with nature and sowed the seeds of their own destruction. They variously triggered soil erosion and the desiccation of the rainforest, or dried out the climate. The lesson we should learn, according to this argument, is that our own efforts at deforestation, which are taking place much further and much faster than any before, hold the same peril in an aggravated form. In other words, like civilizations before us, we are doomed. 'A tragedy is about to repeat itself,' as one journalist put it.

Well, maybe. We cannot dismiss the argument. Many rainforest societies did disappear with a seemingly brutal finality, and the demise may hold lessons for today. But the evidence for some kind of ecological Armageddon is generally rather thin. The Mayan civilization is often described as succumbing to an environmental implosion resulting from deforestation and rising population. But recent evidence suggests that in fact it was a long run of droughts in the ninth century AD that finished it off. Ecological factors may have hastened the decline, but equally the Mayan network of reservoirs and canals and grain stores could have kept the society going for longer than it would otherwise have done.

It might equally be argued that all civilizations eventually come to an end. Why single out ecological factors? Why single out rainforest civilizations? The most unambiguous evidence of crashing rainforest cultures relates to the arrival of human epidemics rather than ecological decline. But in any case, there is something odd about describing societies that lasted for many centuries, even millennia, as having somehow failed, of being 'doomed' from the start. It is equally valid, surely, to argue that, far from being doomed, the societies of Maya, Angkor, Sungbo's Eredo and the rest were success stories of quite remarkable longevity. Maybe we should celebrate their success in managing the forests rather than condemning them for it. Maybe that success offers us equally useful lessons for today.

GOING BANANAS

Around ten thousand years ago, as the Ice Age melted and the Ecuadorians were cultivating the first giant marrows, Stone Age plant breeders were also busy somewhere in southeast Asia. They stumbled on a seedless, and hence edible, mutant form of a wild jungle herb called *Musa acuminate*. They began to take cuttings so that, despite its absence of seeds, they could grow more. They had invented the banana. The new wonder-fruit spread across the region and beyond, first through India and then into the Middle East, where ancient Assyrian wall carvings show it being eaten there seven thousand years ago. The first banana boats crossed the Indian Ocean to sub-Saharan Africa, carrying the delightful fruit that everyone wanted to grow. The mineralized remains of bananas found recently in a fossilized Cameroonian rubbish pit put that journey at at least 2,500 years ago.

In Africa the banana took off in an even more spectacular way. Pierre de Maret, a cultural anthropologist from the University of Brussels, who made the rubbish-pit find, believes the wealth and food security provided by this highly productive rainforest crop probably helped the Bantu people, who were not forest natives, to spread across Africa and eventually become the dominant people across the continent, eclipsing the pygmies and other ancient inhabitants of the forests. The Portuguese quickly spotted the banana when they reached Africa in the fifteenth century and took it to Latin America as early as 1516. By the early nineteenth century a vast profusion of traditional varieties of banana could be found in small plots and forest clearings across the tropics.

By comparison, our modern propagation seems puny. The two main varieties developed for European consumption were both established when colonial botanists brought back promising finds from the gardens of southeast Asia in the nineteenth century. In the 1820s, the French found the Gros Michel, which did very well for more than a century until one of its old Asian pests caught up with it. A soil fungus that produced a wilt called Panama disease started to infest plantations in Latin America forty years ago and, within a few years, effectively wiped out the Gros Michel across the world. Luckily there was an alternative.

Also in the early nineteenth century, an English plant collector called Charles Telfair found bananas in southern China. He sent samples to a friend in England and they were passed on after his death to the Duke of Devonshire, who successfully grew them in glasshouses at his ancestral home of Chatsworth. He named them after his family, the Cavendishes. The Cavendish banana did well enough, but only found world domination after the Gros Michel foundered. Now it is the world's top banana, the only one grown in the great Latin American plantations and the only one you will find in the supermarket. But its fame could prove short-lived. A new version of Panama disease is currently chasing the Cavendish, with potentially disastrous results. As the worldwide head of banana research, Emil Frison of the International Plant Genetic

Resources Institute, put it: 'When this strain of Panama disease reaches Latin America, it will do to the Cavendish what its predecessor did to Gros Michel.' Currently, there is no replacement.

If plant breeders are to find a way out of this bind and discover a future for the world's favourite fruit, they will likely have to go back to the myriad varieties still grown in the tropics. Most are grown not as fruit but as a source of starch. Some 85 per cent of the world's bananas are of these types, grown for frying, mashing, boiling, chipping, steaming, for making banana ketchup and flour and beer and gin. After the three grains – corn, wheat and rice – the banana is the world's fourth most important crop, a staple for half a billion people in Asia and Africa. It is grown on a third of all cultivated land in Uganda, where they eat fifty times more than even British banana-lovers. The banana is the ultimate fruit of the forest, and is really very little changed since it was first plucked from a giant herb somewhere in southeast Asia ten thousand years ago.

BANANA REPUBLICS

The cultivation of the standard banana for global sale began at the end of the nineteenth century, at a time when the rubber boom was at its height. And the same corporate moral climate prevailed, leading to the rise of the so-called 'banana republics' of Central America.

One of the founders of this corporate culture was Minor Cooper Keith, for whom bananas began as a sideline to his main ambition to drive a railroad from the coast into the still-rainforested interior of Costa Rica. His Tropical Trading and Transport Company proved a great success and the bananas had taken over both the Costa Rican rainforest and his business by the time it merged with its main competitor to form United Fruit in 1899.

Banana plantations proliferated across the lowland rainforests of Central America. By the time of his death in 1924, Keith was known as 'the uncrowned king of Central America'. Whole states became, in effect, company towns for the giant US banana corporations. United Fruit became the most notorious of them after organizing a coup in Guatemala in 1954. Keith formed an empire in the jungle – and revealed another. For it was his banana billions that paid for the uncovering and renovation of Mayan ruins in the jungle at Quirigua in eastern Guatemala. Hubris or philanthropy? *National Geographic* in 1913 reported that the Mayan ruins revealed a 'master race' that 'had conquered in easy battle the fever-ridden natives'. In fact, the Mayan master builders were probably themselves locals. But the tag could equally have been applied to United Fruit.

SHAVING THE PLANET

Whatever the strange truth about mankind's past use and abuse of the rainforests, nothing matches the biological holocaust that is now taking place. The whine of the chainsaw rings through almost every jungle; and, once cleared for plantations or cattle pasture, few forests today have a chance to recover. Some of our greatest, most important jungles will be gone in a decade or less. What hope then for the Sumatran tiger or the orang-utans of Borneo? What hope either for the people who depend on the jungle? For some, the hunters are demons – but bushmeat feeds millions. And what fate for all of us as deadly diseases like AIDS and the Ebola virus escape the forest confines and look for new habitats?

TWENTY-FIRST-CENTURY FOREST

It was just one tragedy among many, but it was one that I saw first hand. Loggers had torn through a 'natural laboratory' the size of the Isle of Wight, wrecking the efforts of international researchers to monitor the biology of one of the world's great tropical peat-swamp forests. As hundreds of men with chainsaws forced their way through her domain in Indonesian Borneo, I stood with British biologist Nichola Waldes amid the wreckage of the laboratory's most precious corner, a reserve where all the main local plants had been gathered for research.

'The whole plot is just massacred,' she said. It was littered with broken wood and toppled trunks. Metal tags that once marked every sapling were strewn around. The broken forest was criss-crossed by ramshackle wooden railways and canals, installed by the loggers to help extract logs as they pushed ever deeper into the swamp. On the far bank of the river Sebangau, thousands of logs were lashed together in a giant raft floating beside a sawmill. More logs headed downstream to plywood factories and pulp mills.

This was Central Kalimantan, the most remote province in Indonesian Borneo. It is the size of England but with a population less than that of Essex – and the local phone book lists six times as many sawmills as taxi firms. The natural laboratory is the brainchild of Jack Rieley from the Department of Geography at the University of Nottingham. It is backed by the British Government's Darwin Initiative, a fund dedicated to investigating and protecting biodiversity. The forest here contains an estimated five hundred orang-utans, one of the largest surviving populations of a species under severe threat. But it is disappearing faster than it can be monitored.

The wreckers of Waldes's reserve were not hard to find. As we stood there in the swamp, about a dozen of them paddled into the reserve along a newly dug canal. They came in three canoes, with their lunch in plastic bags and a couple of chainsaws, stopping next to a thick log of red meranti, already chopped down and ready for the sawmill. The leader, who introduced himself as Mr Udin, said they had been in the area – Waldes's natural laboratory – for three months. He claimed not to know that the forest was protected for science. Who was his boss? He shrugged. He knew enough to know they were in the wrong. 'It is dangerous to confront these people. They have police protection,' said Waldes. So we chatted with the loggers over our picnics, as a snake plopped into the water close by. The loggers eventually headed off back down the canal to the river, chainsaws still stowed. Perhaps we had won a small victory for the forest. But they would be back the next day.

For the moment, Central Kalimantan is still the most heavily forested part of Borneo. The forests sit on top of the largest, oldest and deepest tropical peat-swamps in the world. Once, this boggy terrain was ignored: it was too difficult to penetrate and loggers left it alone. But since the late 1990s, when the Indonesian economy went into free-fall, there has been an orgy of illegal logging here. 'There appears to be a conspiracy in Central Kalimantan to extract all the saleable timber as quickly as possible,' said Rieley. Mafia-style organizations have been shipping in thousands of unemployed men from across the country. Often they are men who moved to the cities for jobs but lost them when the Indonesian economic miracle ended.

According to Rieley, groups of up to two hundred men are living in small sections of the forest, felling trees. I saw several teams of Javanese men jammed into trucks driving up and down the new Trans-Kalimantan Highway, which ploughs right through the swamp. They were joining tens of thousands of 'transmigrant' families – whole communities moved by the government from the densely populated island of Java over the past twenty years to the remoter, emptier and more forested islands of Indonesia – the fourth most populous nation on Earth. Many, dumped in an inhospitable land about which they knew nothing, have taken up logging, gold mining and other illegal activities. The British government has put a lot of aid into Central Kalimantan to research ways of making legal forestry more environmentally friendly. But there is little legal logging activity left, Rieley told me. 'These expensive research projects are meaningless without effective control in the forests themselves. There is an air of corruption stretching from the lowest official to the highest reaches of the Ministry of Forestry in Jakarta that is providing business interests with the freedom to pillage.'

* * *

Indonesia stands beside Brazil as the most forested and biologically diverse nation on Earth. Spread across an archipelago of thousands of islands, its rainforests, peat-swamps, mangroves and coral reefs contain a tenth of the world's plant species, and approaching a fifth of its mammals and birds. Pole-axed by an economical collapse in the late 1990s, it has come back fighting – with chainsaws, guns and firebrands. Economic power has shifted from the half-empty tower blocks and factories of Jakarta and its satellite towns on Java to the forests of Kalimantan and Sumatra, the palm-oil plantations on former forest land in Sulawesi and the mines in the jungle interior of Irian Jaya.

Currency devaluation following the crash left imports impossibly expensive, but exports of the country's own natural resources hugely profitable. It has been boom time for the more unscrupulous timber barons and people who want to clear forests to plant palm trees, cocoa and coffee. At the start of the new millennium, tens of millions of city-dwelling poor have been returning to the forests and former forest land. As the slums empty, the forests fill.

Back in the 1990s, the government of Suharto was widely condemned by foreign observers for giving out huge logging concessions in the forests to the president's friends and cronies. Now those days are looked back on with almost fond reflection. The concession-holders at least harvested the forest intelligently enough that they could return in future and harvest more. But now most of the concessions have been invaded by fly-by-night logging gangs, employed by shadowy mafia figures, often with links to the country's military. They have no such concerns.

In Central Kalimantan, I watched as hundreds of workers fresh off the boats from Java filled trucks heading for illegal logging grounds in the swamps. Every few hundred metres along the road out of Palangkaraya, the fly-blown capital of Central Kalimantan province, there were makeshift wooden tracks that are used to slide cut logs over the boggy ground. In his Jakarta office high in a tower block, I met former forestry scientist Tandiono, managing director of a giant Chinese-owned timber concession-holder in Central Kalimantan, who complained that illegal loggers had burnt down his headquarters because he had recruited local villagers to report on incursions into his concession. 'The political and social situation is terrible,' he told me. 'Our company tries to be professional and responsible. We are not perfect, but we try. But the illegal loggers just cut everything, even small trees. The organizers are untouchable. Many of them are army people.' Shortly afterwards, his legal operation shut down.

The pirates are the rule rather than the exception today. The majority of the Indonesian timber harvest, which takes at least 3 million hectares of forest a year, is now illegal.

Indonesia is responsible for a staggering quarter of all the annual disappearance of the world's forests. 'Deforestation on this scale at this speed is unprecedented anywhere, ever,' said Emily Matthews of the Washington-based World Resources Institute. In Brazil, there is enough rainforest left for decades of destruction. Similarly in Central Africa. But here, in the third great rainforest region of the planet, the last of the forests are within a decade of disappearing.

The loggers are after ramin, a tropical hardwood that grows in the swamp forests of Borneo and Sarawak. It is sought after because of its light colour, extreme toughness and straight grain. It is used in everything from snooker cues to picture frames and furniture. Italy, as Europe's biggest centre for quality furniture manufacture, is the biggest importer in Europe. But other hardwoods are cut too. Most of the timber goes to Europe – where much of the lesser wood becomes plywood or pulp for paper manufacturers – or to China, which in the past five years has replaced Japan as the world's biggest timber importer. This followed a decision to shut down its own logging industry because of fears that deforested hillsides were causing landslides and flooding on great rivers like the Yangtze.

According to a World Bank study, virtually all the lowland rainforest of Sulawesi is gone, and at current rates of clearance those of the giant island provinces of Sumatra and Kalimantan will likely be gone by 2010, taking with them the two islands' unique species of orang-utans, tigers and rhinos, as well as thousands of less well-known species. Among the most notable losses will be the Tesso Nilo rainforest on

Sumatra, which has the highest recorded diversity of lowland forest plants in the world, as well as elephants, tigers, gibbons and tapirs.

The wild populations of Asia's only great ape, the orang-utan, which lives only in the forests of Borneo and Sumatra, appear all but doomed. Their numbers have halved in a decade, and even the rainforest reserves set aside for them are being invaded by loggers and palm-oil planters. They include the Tanjung Puting Park in Central Kalimantan, which is home to nine of Borneo's thirteen primate species, including some two thousand orang-utans; and Gunung Leuser in northern Sumatra, where orang-utans live with Sumatran tigers and rhinos, Malayan sun bears and clouded leopards.

Meanwhile, there is growing tension between the outsiders and the indigenous Dayak people of the forest. There have been riots and even massacres. The Dayaks are part of the modern world. Most watch TV and ride motorbikes – but they rely on the natural resources around them to survive, and they say the outsiders are wrecking their heritage. Suwido Limin is a university biologist and a Dayak, living in Palangkaraya. He took me to his family home the night before I left. As we ate coconuts grown on his land, he showed me pictures of carp as big as dogs, which he had matured in his fish ponds. But these were only pictures: he had caught his last fish a year before. The ponds, which once filled with clear spring water from the forest behind his home, had become clogged with silt flowing from a granite mine that opened over the hill a couple of years before. 'This,' he said ruefully, 'is what they are doing to us.'

DRAINING THE PEAT-SWAMPS

At least half of the world's tropical peat-swamps are in Indonesia. The Central Kalimantan peat-swamp forests could be the largest single home for orang-utans, housing more than a third of the fifteen thousand estimated to remain in the wild.

Tropical forested peat-swamps like this one are among the world's most vital biological resources. Not only are they havens of biological diversity, they are also vast repositories of carbon, hugely more so even than the trees that stand above them. The peat in the swamps can be tens of metres thick and as rich in carbon as a coal seam – which many of them, given a few million years, would become. They are soaking up carbon dioxide as effectively as they soak up water. If they are destroyed then the carbon will swill right back into the atmosphere.

Indonesian peat-swamps are about the deepest, most carbon-rich on Earth. By Rieley's estimates, the swamp forests of Borneo hold more carbon than all the world's power stations and car engines emit in four years. The formation of these tropical peat-swamps may have absorbed enough carbon dioxide from the atmosphere to trigger ice ages: to release that carbon now, as the world struggles to counter global warming, seems folly indeed.

It would be nonsense to claim that these peat-swamps are 'virgin' forest. They may be difficult to penetrate, but locals have been hunting and farming in them for millennia. But what is happening now is very different from previous efforts at deforestation, and government agricultural schemes are on a scale probably never seen here before.

In the mid-1990s, Indonesian President Suharto decreed that an area of Central Kalimantan peat-swamp forest should be turned into a giant rice paddy, making his country self-sufficient in its staple foodstuff. Suharto brushed aside scientific advice that the scheme, involving 4,000 kilometres of canals to drain the swamp and divert rivers, would be an ecological and economic catastrophe. Some 100,000 transmigrants moved on to the swamps. But, says Gunawan Satarl, the head of a government scientific review of the scheme, it 'destroyed 1.4 million hectares of forests, caused forest fires and polluted rivers', before the incoming President Habibie abandoned construction in 1999. Meanwhile, hundreds of local people, far from welcoming this investment, have been demonstrating about the loss of rubber, timber and rattan stands, and fish that they used to catch in ponds as the swamp flood waters retreated. Patrice Levang from the French government development research institute Orstrom, who has advised the Indonesians on transmigration for twenty years, told me: 'The results are even worse than the worst forecasts. Once you drain these areas you can no longer control the water. Some places are flooded and their plants have drowned.' Elsewhere the canals for the irrigation project have dried up the swamps, leaving them desiccated and friable. And, in the event, the peat soils proved useless for growing rice. When Suharto came in 1998 – just before his downfall – to ceremonially harvest the first rice crop, nothing had grown. So officials transplanted rice from elsewhere to fool him.

RAINFOREST PERSPECTIVES

Indonesia is the current hub of global rainforest destruction, but it is far from being the first in the modern era. Much of Central America was cleared for banana and other fruit plantations a century ago. Huge areas of West African jungle were cleared in the late nineteenth and early twentieth centuries, largely as a result of imperial adventures in logging and plantation agriculture. The British colony of Sierra Leone was typical. Vivien Gornitz of Columbia University in 1985 summed up: 'Most of Sierra Leone was formerly covered by evergreen and semi-deciduous forest . . . The widespread deforestation began with the settlement of Freetown [the capital] and environs in the early nineteenth century. Logging for teak and other hardwoods began along the rivers of northern Sierra Leone, expanding southwards by the 1840s. Old palms invaded the original forest, though the interior remained largely forested until early this century. But, by 1947, the mountains were bare and the plains largely grassland, with only patches of secondary thickets. Coastal mangroves have been cleared for rice production, partly to compensate for the severe erosion caused by upland rice growing earlier in the century.'

After the Second World War, the focus of deforestation shifted to Asia, where the timber industry ransacked each forested country in turn, from the Philippines to Thailand and Malaysia, each of which have lost most of their large stretches of rainforests in the past half-century. Now it is

Indonesia's turn. Until recently the biggest market was Japan, but today it is China.

But logging is far from being the only pressure on tropical rainforests. In some areas, such as the Brazilian Amazon, clearance of land for cattle pastures is the biggest threat. In others it is industrial-scale plantation agriculture, especially those growing palm-oil and soya. Elsewhere it is pressure from landless farmers, whether through organized migrations such as Indonesia's long-standing transmigration policy, or piecemeal movements by poor farmers, or the occasional frenzies caused by discoveries of gold. Always in the background lie the politics of land. Rainforests in most developing countries are the final frontier, where the poor and dispossessed, the rich and omnipotent, or the plain criminal can all attempt to fulfil their dreams.

Nobody is sure how extensive tropical forests once were. One widely accepted estimate is that they covered 25 million square kilometres – 16 per cent of the land surface of the planet – though, as we have seen, such estimates have to be handled with care. Today, they are below 10 million square kilometres. Half of the loss has been to permanent farms, another quarter to pastures and the final quarter is at any one time under shifting cultivation. It is arguable to what extent these forests are 'lost', since the process of shifting cultivation has been going on for thousands of years, in most cases enabling the forest to recover after the farmers have moved on. Whatever humans got up to in the past, the pressures are greater today and the massive deforestation of the great tropical rainforests today is real enough.

As plantations and pastures, roads and pipelines, mines and logging camps spread, the jungle is being broken up into small fragments. Even great forest regions like the Amazon and the Congo are becoming a series of small islands; and, as that happens, biodiversity begins to slide alarmingly. A study of the growing number of 'forest fragments' in the Amazon shows that small chunks of rainforest preserve many fewer species than larger ones.

Researchers from the US and Brazil's National Institute for Amazonian Research in Manaus spent thirteen years catching birds in patches of undisturbed forest that had been isolated by clearance for cattle pastures. During the study, the number of bird species declined in all the fragments. But the 'half-life' of the biodiversity – the number of years it took for half the species to disappear – was dramatically shorter in the small patches. 'Fragments of 100 hectares lost half their species in less than fifteen years,' said Goncalo Ferraz of Columbia University in New York, who led the study. Bigger fragments lost progressively fewer species, but even a 100,000-hectare fragment lost 5 per cent of species in fifteen years.

Conservationists are increasingly trying to protect the remaining rainforests by creating protected areas that link up small surviving patches so new forests can form in the gaps. But the new findings suggest that many such projects are doomed: unless the original fragments are big enough, then the species the conservationists are trying to protect will disappear in the twenty to fifty years it takes for the interlinking forest corridors to re-grow.

Of course survival rates depend to some extent on the

species involved. Big predators like jaguars and tigers need more land but can often move between the patches to find their next meal, so long as there are no human hunters around. But other species will not. For practical purposes, says Ferraz, trying to preserve rainforest fragments smaller than about 1,000 hectares, or 10 square kilometres, will not work.

But the fact that rainforests are fast disappearing and facing lethal fragmentation in some areas does not mean they are disappearing fast everywhere. In Central Africa today, they may still be close to their maximum extent for thousands of years. In early 2004, when forest scientists began to return to the more remote forests of the Congo after a decade and more of civil war, many expected to find carnage. They feared that rebel fighters and government armies would have used the war to rampage through the jungle, killing its wildlife and setting up clandestine logging concessions. They did find some evidence of lawlessness: troops from Zimbabwe, ostensibly there to protect the administration, had been running logging operations in parts of the country. But when Gottfried Hohmann of the Max Planck Institute for Evolutionary Anthropology in Leipzig carried out a survey of the Salonga, Congo's largest national park, he found to his amazement that war had in fact helped protect the forest. 'Salonga has survived the war in excellent shape, with 98 per cent pristine forest,' he said. What he did fear was the peace. In the week that he published his results, Congolese president Joseph Kabila was touring Europe calling for logging companies to invest in his now-peaceful country.

GLOBALIZATION, BEEF AND PALM OIL

The globalization of trade is inevitably a threat to many rainforests. Clearing land for cattle pastures is the biggest single driver of rainforest destruction in the Amazon. And most of the beef is today produced for export. By the year 2003, Brazil had become the world's biggest source of exported beef, and most of it came from former rainforest regions of the Amazon. 'Brazilian deforestation rates are skyrocketing and beef production for export is to blame,' reported David Kaimowitz, director-general of the Center for International Forestry Research in early 2004. 'Until 1991, ranchers in the Amazon used to sell their beef only in that region; then it was only in Brazil, but now they have access to the entire world market.' Exports had increased fivefold between 1997 and 2003, with four-fifths of the increase coming from the Amazon. Brazil was exporting to Europe, Egypt, the US, Russia, Saudi Arabia and many other nations.

Weeks after Kaimowitz's report, Brazil released data showing that deforestation rates in the Amazon had risen by 40 per cent in the previous two years and was higher than for any preceding twenty-four-month period. Some 50,000 square kilometres had been razed, the equivalent of eleven football pitches every minute. The Amazon states of Rondonia, Para and Mato Grosso had seen both the biggest increase in cattle-herd size and the greatest rates of forest loss. Ranchers, penetrating remote rainforest along government-built roads, were occupying forest land nominally owned by the government, felling it and bringing in cattle. Interestingly, native Indian communities were far better at defending their land than the government. Ranches often ran right up to the edge of the lands of indigenous peoples, but did not usually cross over.

While the livestock industry is driving wholesale forest removal in the Amazon, in Indonesia the logged-out former forest land is mostly being converted to palm-oil production. The oil palm, which originates from West Africa, grows up to 15 metres tall and is the world's highest-yielding vegetable oil crop. We use it to make biscuits and ice cream, cooking oil and margarine, cosmetics and soap. By one estimate a third of all foods on supermarket shelves contain palm oil. Indonesia is trying to overtake Malaysia as the world's leading producer of palm oil. Most of the forests lost in Sumatra during the late 1990s were converted to oil-palm plantations.

MEAT FROM THE BUSH

Gregory Etim Inyang hunts elephants because he can see their eyes in the dark of the forest night. But his friend Edet Okon says it is difficult making a kill because 'they run away even when they have been shot'. Monkeys, he says, are good prey. But you have to be a good shot. Both men are agreed, however, that hunting near their home in the last surviving patches of rainforest in southeastern Nigeria is a loser's game. 'The forest is almost finished,' says Edet. He has to travel too far before he can find a catch, whether it is bushbuck or pig, buffalo or monkey. 'It's a five-hour trip,' he says. 'Eight hours,' says Clement Inyang. 'The animals are getting very scarce.'

All these hunters live in a single village, Ekonganaku, not far from the Cross River National Park, whose forests are among the last surviving homes for large wild animals in Africa's most populous nation. Their stories emerge from interviews conducted by John Fa and Sarah Seymour from the Durrell Wildlife Conservation Trust on the Channel Island of Jersey. Fa is one of the world's experts on bushmeat and those who hunt it. He wants to know how and why forest dwellers still hunt – and, most crucially perhaps, whether their children, in thirty years' time, will still be out in the bush with traps and guns.

Many hunters seem in little doubt as they head out into the jungle that they are engaged in the last round-up. 'The animals are finished because there are too many hunters,' says Anbor Inyang, who specializes in killing monkeys. 'Once there were many animals,' says Felix Edet Etim, who loves the taste of porcupine, when he can get it. 'No, pangolins are best,' says Paulinus Ayuk, 'very sweet.' Another prefers chevrotain, a small hornless deer, for its 'colour, beauty and tenderness'. But this is lost on Paulinus, who says, 'I catch anything for money.'

'They recognize what is going on,' says Fa. 'They know the bushmeat business is no longer sustainable, that there are too many hunters, that the big animals are mostly gone from round the villages and that they have to go further and further to find them.' Even when the jungles survive, they are often becoming empty of large animals. The hunters used to regard wild animals as an infinite resource, but nobody sees it that way now. 'They don't need educating about the problem,' Fa says in a side-swipe against conventional environmentalism. 'They don't need heavy-handed policing. They need alternatives – to things to eat and to the means of making a living. Trying to stop them from hunting won't work otherwise.'

This debate matters not just for the wildlife, but also for the people of Africa. In many parts, particularly of Central Africa, where the continent's biggest rainforests survive, bushmeat makes up 80 per cent of the population's protein intake. And it is not just deep in the jungle or among poor villages that bushmeat is the norm. Go to the supermarket in many Central or West African cities and you can pick up a shrink-wrapped pack of elephant meat – it was probably brought out from the jungle on logging trucks. And bushmeat is becoming fashionable, like organically grown food in Europe. In Burundi, the dish of choice in some of the best restaurants is hippo steak. Through Ghana, small boys stand by the roadside dangling giant rats. Drivers stop to buy supper as routinely as if they were dropping into the 7-Eleven on the way home from the office. If there are risks of disease from this jungle meat, nobody is concerned about it.

* * *

Animal rights activists and many environmentalists are up in arms about the bushmeat business. As part of their propaganda war against the hunters, they send out gruesome images of gorilla corpses and monkeys strung from poles, illustrating literature on 'the slaughter of the apes'. A particularly unsettling image, caught by the anti-bushmeat campaigner Karl Ammann, shows a freshly cut gorilla head in a kitchen bowl on a sideboard, next to a bunch of bananas. Such reports concentrate entirely on the burgeoning bushmeat business as a crisis for wildlife and Africa's biodiversity. There is no question that some of the world's most endangered wild animals are, quite literally, being eaten to death. Dozens of species of great apes, monkeys, snakes, rodents, deer, birds and maybe even hippos face extinction in the tropical rainforests of Central Africa as a result of being hunted for food. Many in Asia have already gone. And the growing bushmeat trade is now seen by many conservationists as the single biggest threat to the existence of many rare animals – more important even than the destruction of their habitat through deforestation.

But, asks Fa, is our Western squeamishness getting in the way of a sensible appraisal of the importance of bushmeat?

Are we in danger of caring more about the survival of a few rare rainforest species than with the survival of their hunters? If we condemn all hunting for bushmeat, then how do we propose that Africans eat? And equally, if we do not take the trouble to find out why bushmeat hunting is so prevalent still in Africa, how can we hope to stop it?

Fa is the first researcher to have attempted to quantify systematically what is going on. In a series of investigations in West and Central Africa in the past decade he has charted the hunt, measured its importance for nutrition and watched how the types of animals showing up in market places across the continent have changed. Extinctions of many important and well-known jungle species, such as chimpanzees and the great apes, will begin 'relatively soon', he says. They will be widespread within a few decades. The rainforests will become 'ghost jungle'. In fact, in the dwindling rainforests of West Africa, the first modern primate victim has already been hunted to extinction. In late 2000, scientists reported the passing of Miss Waldron's Red Colobus. This habitué of the rainforest canopies of Ghana and the Ivory Coast had been brought down by bushmeat hunting. It was the only primate species to become extinct in the twentieth century but the twenty-first century is likely to see many more go the same way.

Bushmeat has always been part of the staple diet of forest dwellers, though, contrary to perceptions in the West, they are not great meat eaters. 'People don't actually eat a lot of meat day today,' says Fa. 'They use small amounts to flavour soup, or they sell the meat on for cash. The exception is during festivals, when young men will go out to catch meat to show off their prowess.' But, again contrary to many people's perceptions, rainforests are not and probably never have been teeming with wildlife. You can walk for days without seeing wildlife unless you know how to look. The biological productivity of a typical piece of rainforest is typically less than a tenth that of open bush, says Eleanor Milner-Gulland of the Renewable Resources Assessment Group at Imperial College in London.

The central problem in the African rainforest is that as the human population soars and cities grow, national economies have failed to produce domesticated meat of the kind familiar in stores in more developed countries, or even in equivalent cities on the edges of the Amazon or southeast Asian rainforests. Much of the continent, urban as well as rural, still relies on the bush for protein. They eat wild elephant rather than pork; snake rather than chicken.

* * *

Hunting takes place almost everywhere, even in many national parks, which ought to be refuges where animal populations can recover, undisturbed by trap or gun. Village hunters who once killed solely for the pot now also kill for cash. And with firearms widely available after so many civil wars, hunting is no longer a specialist skill. There are few full-time hunters, but millions of farmers go out hunting when they have little to do on their land. 'Hunting is very seasonal,' says Sarah Seymour. 'Whenever people are not harvesting they will hunt, and they will take a gun whenever they go into the forest.' Many of these part-time hunters are pretty indiscriminate, she says. 'Often, they shoot first and then see what they have bagged.' They sell at the roadside or to traders heading for nearby bushmeat markets in the towns. Even children do it. In Cameroon I have spoken to children who have caught monkeys almost as big as themselves using slings, and carry dead and wounded animals round looking for a sale.

Working in their small region of Nigeria close to the border with Cameroon, Fa's team has produced the most detailed study yet of bushmeat hunting and trading. 'We have data on about 100,000 transactions of animal carcasses in almost a hundred villages,' he says. Ekonganaku's local market for bushmeat is at the village of Ningajue. It is no big deal. It is open for business once a week, on a Saturday, for a couple of hours from 6 a.m. Some hunters deliver directly to the women running the market, arriving with meat in large sacks. But others from further away, including those from Ekonganaku, sell to traders with motorbikes or taxis, who bring the meat to the market. 'There are still lots of species sold at Ningajue,' says Seymour. 'There are usually about a dozen sellers, each with maybe fifteen carcasses. I see grasscutters and rodents, chimpanzees, red colobus and duiker, a type of antelope. But the most popular meat is bush pig, which is still quite plentiful and very tasty.' There are other species available under the counter, she says. 'But if they have an illegal species I have to leave because they think I might recognize it.'

In bigger towns, such as the regional centres of Calabar, the bushmeat market is daily. And much meat goes directly to big customers like restaurants and canteens, without ever crossing a market stall. All down the supply chain, prices rise. A monkey carcass in Ekonganaku costs four hundred naira, or two pounds, but it fetches a thousand naira, approaching five pounds, in Calabar. In the big markets, monkeys are prepared, their heads, hands and feet cut off, and smoked. 'You would only know it was a monkey, not which kind or whether it was one that could be hunted legally,' says Seymour.

As the animals in the Nigerian forests round the villages give out, hunters are travelling further, defying the occasional guard to go into the nearby Cross River National Park, for instance. And more and more meat is being brought across the border from Cameroon; this is partly because there are still more animals in the forests of Cameroon, especially in the Korup National Park, close to the border, where efforts by WWF and others to prevent hunting have largely failed. But it is partly because prices are higher in Nigeria, so Cameroonians prefer to export. In any event, the Cameroon forests are emptying. 'In ten years, Cameroon will be like Nigeria today,' says Seymour.

Fa has carried out similar studies in Equatorial Guinea. There, he tracked the activities of forty-two hunters for more than a year, during which time they had set more than 100 million snares and caught 3,000 animals, or some 11 tonnes of meat. 'It's very well-organized,' says Fa, 'a wonderfully well-run commercial enterprise. Taxis bring in the meat from the bush to the main market at Malabo each morning. But what you often see is that there isn't that much meat.' For all the effort, the market sells only about twenty carcasses a day.

'That's less than a rack of meat in a small supermarket. The bottom line is that for all the effort, the results are pitiful. There is just not enough meat left in the forests.'

It is almost always the big animals – the ones we know and love – that die out first. Fa's Malabo study was a repeat of a similar exercise five years before. During that time he found that monkeys and duikers had been largely replaced with rodents and other small creatures. This switch, which is widespread across Africa, is happening partly because big animals, having more meat on them, are hunted more. But they are also more vulnerable, because large species tend to mature later and have lower rates of reproduction. So they are less able to replace the numbers killed by man. In the jungle, big means vulnerable, says Milner-Gulland. 'Large species like tapirs and primates disappear first and as they vanish people turn to the smaller ones, such as squirrels and cane rats.'

In Central Africa, the animals most at risk of disappearing are the great apes, forest baboons such as the mandrill, and duikers. But there is also growing concern for the fate of elephants and hippos, both of which are often being taken these days as much for their meat as their ivory. When researchers returned to the Virunga National Park in the Congo in late 2003, after a decade of being kept out by armed bands, they found that the world's largest population of hippos had crashed by 95 per cent, leaving the giant creature as the latest of the planet's megafauna in danger of extinction. In the 1970s, some 30,000 hippos lazed in the rivers and marshes of the park. But in 2003, just 1,300 remained. The census, carried out by the Congolese Institute for the Conservation of

Nature, concluded that the major problem was hunters intent on bagging the animal's meat. Hippo meat has become a delicacy in Central African restaurants, selling wholesale for two dollars a kilogram at the Burundian border town of Gutumba. That makes an entire carcass worth three thousand dollars. In Burundi itself, two-thirds of the national population – representing two hundred animals – has disappeared in five years.

There are few official statistics on the bushmeat trade. Much of it, after all, is illegal. But, legal or not, the trade is increasingly commercialized, and there is also a growing cross-border market, even in remote areas. Cameroon meat crosses into Nigeria on foot. Convoys of bicycles carry tonnes of smoked elephant meat down forest tracks from the Democratic Republic of the Congo into the neighbouring Central African Republic, where it turns up on supermarket shelves. And there is a growing intercontinental market. Smoked monkey meat, in particular, is smuggled in suitcases aboard aircraft to the high-priced markets of Europe. Dalston market in London's East End emerged in 2002 as an important outlet. It is clear from this, says Fa, that bushmeat is often a luxury item of such cultural significance to people of recent African descent that even people who have not lived on the continent for decades still seek it out.

* * *

The biggest centre on the planet for bushmeat hunting is the Congo basin of Central Africa, where most of the population still lives in villages in the bush. The basin's rural population is around 34 million people. And there are millions more in cities surrounding the forested basin. A typical square kilometre of Congolese rainforest today contains around twenty people, most of whom get their protein from hunting the animals living close by. Fa reckons around 5 million tonnes of bushmeat is caught and eaten here each year. It is equivalent to a staggering 90 per cent of the weight of all the animals estimated to be living in the forest at any one time. Put simply, if the animals did not reproduce, the forest would have been empty in thirteen months.

Luckily they do reproduce. And reproduction has kept the populations of most species limping along. But this cannot continue. Fa calculates that hunting is currently being carried out at between twice and five times the rate at which animal populations can reproduce without facing eventual extinction. With human populations still rising, and demand for bushmeat still on the increase, the days of the final screech of a monkey or scuttle of a bush pig through the rainforest may not be long in coming.

Hunting is becoming just another economic activity in the jungle. Many Central African forests are being invaded by miners and logging gangs, often working for French, Belgian or other European companies. Often, these gangs are not supplied with food, but are given guns and told to go out and hunt for their supper. Locally such gangs can have a devastating impact on wildlife populations; and the roads built to remove the timber and minerals also make transport of animal carcasses to distant cities far easier. Bushmeat hunting becomes a lucrative sideline for the gangs, and truckloads of bushmeat are unloaded at city markets each morning.

The popularity of bushmeat is not surprising. A study by the environment group TRAFFIC, which investigates the trade in endangered species, found that right across Africa bushmeat is generally cheaper and more available than domesticated meat and that 'the poorer households are more greatly reliant on bushmeat'. But the rich also seem to be partial. Until early 2003, when the Cameroon government announced a ban on serving endangered animals, many upmarket restaurant menus in Yaoundé, the capital, included gorilla, chimpanzee and elephant.

Fa used dietary studies to estimate just how much bushmeat protein Africans consume. Figures ranged from 28 grams a day in Congo, the poorest country in Central Africa, to 180 grams in Gabon, the richest. The UN recommended minimum intake is 52 grams. So it is clear that, in many areas, bushmeat is an essential part of a subsistence diet. Perhaps if the rich backed off bushmeat and hunters stopped hunting for cash, the situation would be better. But, as Milner-Gulland points out, it is also the poorest that rely most on commercial hunting. 'They are the most remote and marginalized groups, who have few easily available alternative sources of cash.' The commercial bushmeat business is essential both to feed millions of city dwellers and to provide cash for the poor in the bush. And, as things stand, hunger looms as the forests empty. Fa estimates that, by 2050, the Central African bush will be able to supply just 9 grams per head per day for the region's population. Unless another source of protein shows up, 'malnutrition is likely to increase dramatically'. The only other source would be domesticated meat.

Fa's research has led him to come out against the conventional environmental response to the slaughter of wildlife – demands for bans on hunting and trade. He says that environmentalists are in danger of behaving like Marie-Antoinette, who, on being told the French peasants had no bread, replied: 'Let them eat cake.' Pork and chicken are no more available in Central African supermarkets than cake was in pre-revolutionary Paris. Right now, the majority of poor Africans have no alternative to hunting and eating bushmeat. Many countries across the continent are going backwards economically. They have virtually no governments, deteriorating infrastructure and near-constant civil wars. Food production on farms in Central Africa has not risen since the 1960s, and in many areas it has fallen back sharply. Their economies are going 'back to the bush'. How can their diets do other than follow? And of course the one thing that the civil wars do provide them with is guns and other weapons with which to go hunting.

'We are understandably horrified by wild animals, especially primates, being killed for food,' Fa says. 'But we must remember that bushmeat is a cheap source of protein for many malnourished people in Africa.' And only by saving both can the forest be saved.

ASIAN LUNCH

If the bushmeat bar is generally the fast-food outlet of Africa, in Asia wild protein is an expensive delicacy. And as people get richer, they want more of it. Bushmeat is often highly prized in Asia for its medicinal as well as its culinary qualities. Vast markets exist across the continent, mostly supplied from its dwindling rainforests. The trade is insatiable. More than 1,000 tonnes of turtles are exported annually from the forest swamps of the Indonesian island of Sumatra. On nearby Sulawesi, a single food market sells the carcasses of eighty thousand forest rats a year. In the past forty years, twelve large vertebrate species have gone extinct in Vietnam alone, all largely because of hunting. Out in the swamps of Kalimantan, forests are burnt just to root out turtles.

This reptile rush is at its most intense in the Middle Mahakam swamps of East Kalimantan, where traders have been buying up as many reticulated pythons, monitor lizards, banded swamp snakes and freshwater turtles as villagers can supply. A kilogram of soft-shell turtles fetches about one pound sterling for the villagers – more than a day's pay. In one small area around the village of Enggalam in East Kalimantan, WWF's Paul Jepson logged the sale of 143 tonnes of turtles and 19,000 skins of banded swamp snakes in nine months. The hunters dispatch the live turtles to up-market restaurants in Hong Kong, Singapore and Taipei. Many of the snakes are turned into skins and meat locally. But the gall bladders of pythons are particularly valued in Chinese pharmacies.

China has few of its own rainforests left. Just a few small patches remain near its southern borders with Vietnam and Laos. But that has not stemmed the huge market for jungle produce. At the traditional medicines market in Chengdu, capital of Sichuan province, a giant hangar, the size of a couple of soccer pitches, is packed with stalls selling plant and animal parts of almost every description. There are all manner of parts of forest deer, including feet and tails, horns and penises, newts to treat stomach acid and dried snake skins for dunking in wine as a tonic. There are bat faeces, pangolin scales, the fallopian tubes of frogs, toad venom, dried geckos and forest leeches arranged in rows like chocolate bars on a counter.

This market is but one outpost in a fast-growing trade supplying China's apparently insatiable demand for wild medicines from its tropical neighbours. Guo Yinfeng, author of a Chinese government report on trade in endangered species, polled thirteen medicinal-foods manufacturers and found that between them they consumed annually 16,000 tonnes of rat snakes, 200 tonnes of pangolin scales, 25 tonnes of leopard bones, 500 tonnes of mixed scorpions and 6 million geckos. And, unlike medicines in the West, where demand is strictly limited by prescribed doses, in China the distinction between food and medicine is vague and often non-existent. More is better. The richer and more urbanized the Chinese get, the more of such foods they eat.

MANATEES AND MORALS

The Amazon manatee, a remote relative of the whale, is the largest of the animals that live in the greatest rainforest on Earth. This denizen of the flooded forests on the Rio Negro was, like the great whale, hunted almost to extinction in the mid-twentieth century to provide oil boiled down from its blubber. The Brazilian government eventually reacted by providing legal protection for the animal; but forest dwellers still hunt it for its meat. And here is the interesting bit: WWF conservationists backed them – and, indeed, encourage limited hunting in the protected areas that they run.

José Marcio Ayres, who managed the Mamiraua Ecological Reserve in the heart of the flooded forests, wanted to conserve the manatee as much as anyone; but he persuaded the Brazilian government to allow forest dwellers to continue their hunt. He also encouraged the hunters to develop their own quota system to ensure that this valuable food would not become extinct. 'Bans have been tried in many reserves in the Amazon,' he told me on a crackling phone line from the jungle, 'but they don't work.' He also told me, early in 2003, shortly before he died, 'The meat is delicious.'

So, in a project supported by both WWF and the Wildlife Conservation Society, Ayres declared that forest dwellers should have control of their own resource. 'Hunting bans make conservation an enemy of the people,' says Sandra Charity of WWF Brazil, who worked on the project with Ayres. 'Our strategy is to work democratically with the villagers to help them protect their resources.' Ayres and Charity say that manatee numbers fell while hunting was banned by government diktat, but they have revived since the local hunters themselves took control.

Arguably such strategies have a better chance of working in the Amazon, where the human population is much lower than in the Congo basin. Fa calculates that general hunting rates in the Amazon are less than one twentieth of those in Africa. There may be over-exploitation in the vicinity of Amazon hunting villages, but, across the basin as a whole, few if any species are likely to be vulnerable as a result of the bushmeat business, he says. Even so, there is a principle being raised here that goes beyond mere pragmatism. WWF in particular, under its international director Claude Martin, has not been afraid to think the unthinkable in wondering whether environmentalism should go hand in hand with animal rights or with human rights. 'If poor people can't make a living from nature's forests and animals, then they will probably eliminate them with guns or chainsaws to free more land for farming,' says Martin.

An old hand at national park management in Africa, Martin is not squeamish about confronting one of the gravest charges against Western environmentalists – that they are latter-day green imperialists. Dogmatic environmentalism, he agrees, is 'in some ways as narrow and selfish as the imperialism of old. Imperialism imposed a system of development that took little or no account of the rights and needs of local people. Too often that same charge can be levelled against conservation projects.'

DISEASE AND DEFORESTATION

The first European expedition up the Congo, under Captain James Tuckey in 1816, did well while sailing up the river through mangrove swamps and past passive villages. Then the crew took to the land to pass some rapids. Within days, Tuckey and most of his officers had gone down with malaria and yellow fever.

Tuckey's tongue turned first white, then yellow, then brown and finally was coated with a black crust. He suffered headaches that spread to engulf his body in pain, and he died. Less than a decade later, in West Africa, a brawny, red-bearded Scot called Hugh Clapperton, along with his officers and crew, succumbed similarly when they took a short overland rainforest journey from their ships moored off Lagos to the arid interior, in their search for the origins of the river Niger. Nine men left the ship and only one returned, thanks to malarial fever and dysentery.

The great scientists suffered similarly. Darwin got Chagas' Disease; Wallace suffered from repeated bouts of malaria, while his brother and many of their friends in Belem died of yellow fever. Spruce lost his health to the Amazon. A mixture of dysentery and malaria got Livingstone some years after his long-suffering wife – alone in a mission station while he tramped the continent – succumbed to malaria. The rainforests have always been feared as much for the disease they harbour as anything else. Countless expeditions foundered as their members went down with fevers. Disease certainly killed far more explorers than the spear- and poison-dart-wielding natives. For centuries, yellow fever was regarded as the prime reason why Europeans never quite gained control of their West African empires. As doctors told the traveller Mary Kingsley before she headed for Sierra Leone, it was 'the deadliest spot on Earth . . . the white man's grave, you know'. (Mary, being made of tougher stuff than many Victorian men, related in her 1897 book *Travels in West Africa* that 'naturally, while my higher intelligence was taken up with attending to these statements, my mind got set on going, and I had to go'.) Africans, though brought down by malaria in large numbers, saw such diseases as one of the best defences they had against European invaders. As one playground song in Nigeria had it: 'Only malaria can save Africa/Only yellow fever can save Africa/Only mosquito can save Africa.'

Today, more and more humans are coming into contact with jungle diseases. For one thing, more and more people in developing countries are becoming personally acquainted with the forests, whether passing through along roads, eating bushmeat, or joining logging gangs and mining enterprises. For another, some of the diseases themselves seem to be flourishing as the forests are uprooted, especially those that spread via people and their animals, such as cattle. Some of the worst epidemics are occurring among those who live on the fringes of the forests. This deadly crossroads is where mosquitoes, tsetse flies and other disease-carrying insects are most abundant. They breed in the pools of water that form on the ground when the forest canopy has been removed.

People living close to former forests are most vulnerable to malaria in particular. A disease thought to be on the verge of elimination half a century ago now kills up to 3 million people a year, mostly young children.

There are many reasons for the revival of malaria, including the collapse of public-health programmes and the withdrawal of insecticides like DDT. But environmental changes are important, too. Global warming helps, but probably more important is the rise in temperatures as the forests, with their great powers to shade the Earth, are removed. Malarial mosquitoes like best the open, moist and well-vegetated places that people make as they tear down forests and begin farming. When researchers from Johns Hopkins University in Maryland went counting mosquitoes along a jungle road in northeast Peru, they found that every 1-per-cent loss of forest cover produced an 8-per-cent increase in the number of malarial mosquitoes. 'This is happening because the mosquitoes thrive in open spaces,' said Jonathan Patz. Deforestation creates ideal conditions for the spread of mosquitoes by creating a proliferation of warm, sunlit pools of water where they lay their eggs. Once a third of the forest is gone, mosquitoes are in their idea of heaven, Patz believes.

Other mosquito-borne diseases are spreading for similar reasons, including dengue fever in Latin America. But it is not just mosquitoes. In 2004, a mysterious outbreak of rabies began killing dozens of people around the mouth of the river Amazon. Scientists finally linked it to an unprecedented increase in attacks on humans by vampire bats that were infected with the disease. They blamed the attacks on deforestation, which had forced the bats to leave the forests, and on the spread of livestock farming into deforested areas, which increased the food available for the bats. Meanwhile, in West Africa, an estimated quarter of a million people each year catch Lassa fever – a disease that was unknown to doctors until 1969. People catch it from forest rats moving into their houses. In some areas of Sierra Leone and Liberia, up to one in six people admitted to hospitals have the viral disease, and there are thousands of deaths.

* * *

Deforestation, say doctors, is increasingly linked to the emergence of deadly new diseases. The Ebola virus comes out of the Central African jungle – and it keeps coming out. One of the most deadly diseases known to man, Ebola kills around 90 per cent of all the people who catch it, usually within a few days. There is no cure or vaccine. People simply bleed from every orifice of their bodies until they die; and anyone coming into contact with their blood or other bodily fluids is likely to become infected and join them in the mortuary. The only thing that seems to stop it becoming a major epidemic killing millions is that people who catch it die so quickly they scarcely have time to pass it on except to a few family members and the occasional unlucky doctor or nurse. So far, every epidemic has fizzled out; but the first time that a victim makes it out of the jungle and into a crowded city before dying, we could be in serious trouble.

Humans catch Ebola from forest apes, with whom we share a range of diseases. Apparently it is most easily transmitted through eating ape-meat, an increasingly popular activity in Central Africa. A recent outbreak began in December 2002, when six gorillas from a single family group were found dead from Ebola in a gorilla sanctuary at Lossi in Congo, near the border with Gabon. Soon a group of 143 gorillas at the sanctuary had been reduced to seven. And within a month the disease had passed to local humans. The area around Lossi is a rich bushmeat-hunting area, with a brisk cross-border meat trade along jungle tracks. Within weeks of the outbreak among the gorillas, people were dying in villages within a day's walk of Lossi. By early 2003, more than a hundred humans had died. Researchers eventually tracked the origin of the disease in humans to hunters eating a wild boar they had found dead in the jungle.

According to the World Health Organization, there have now been a dozen outbreaks. A detailed study of Ebola outbreaks in humans published in 2004 found that almost all were preceded by the discovery of large numbers of dead apes in the bush nearby. 'The virus doesn't so much spread as spill over from the forest,' says William Karesh of the Wildlife Conservation Society. Left hanging in the air was the question of whether it was spilling over more often because the forest was diminishing. A more immediate question is where the Ebola virus (which is named after the Congolese river where it was first identified) goes between human epidemics. It was once thought that apes harbour it. But, like humans, they die so fast that this is now thought unlikely. Forest birds are one likely reservoir, but others suggest some species of bat or mouse.

Equally unknown is whether Ebola is an ancient disease like malaria, or a modern disease like AIDS. Some believe it has always lingered in the Central African forests and may have been responsible for great population crashes of the past. In modern times, it was first spotted by doctors in the remote border lands between Sudan and the Congo in the mid-1970s. Some two thousand people have died so far: the biggest outbreak to date was in 1995, when more than two hundred died – about a third of them health-care workers.

But perhaps the most dangerous breakout came in late 2000, when Ebola jumped into the human population of western Uganda, a densely populated rural area less than a day's drive from the nearest international airport. As the death toll approached fifty, Ugandan police tried to cut the area off from the rest of the country. But still the epidemic spread and more than 150 people eventually died. The hero of the hour proved to be a doctor, Matthew Lukwiya, who ran the local hospital in the town of Gulu. He stayed to contain the disease – and died of it, after a crazed patient an hour from death sprayed his face with blood. At Lukwiya's funeral, pallbearers wore face masks, latex gloves and surgical gowns to carry the tightly sealed coffin. But his swift isolation of patients had probably prevented something much worse. Will the world be so lucky next time?

There are other, similar viruses coming out of the jungle. One is known as the Marburg virus, whose first known victims were workers at an animal research laboratory in

Marburg, Germany. They were taken ill after working with a consignment of wild African green monkeys imported from the forests of northwest Uganda where, possibly coincidentally, Ebola virus has also shown up. More than thirty researchers became sick, bleeding profusely from all orifices. Seven died. The others suffered permanent damage to their livers and became impotent. Then the disease, which had apparently worked its way through Ugandan monkey populations five years before, disappeared – until a young Australian hitchhiker died of the same virus eight years later, after sleeping outdoors in the bush in Zimbabwe. His girlfriend was narrowly saved. Where is Marburg now? Is it residing in some population of monkeys or other animals somewhere in the African bush? We simply do not know.

* * *

Two-thirds of the thirty new diseases afflicting humans in the past half century have come from animals, many the inhabitants of tropical rainforests. Remember the SARS epidemic of 2003? That began after infection from the cat-like masked palm civet. The cat, a delicacy in China, is not troubled by the virus; but around one thousand humans died before a global effort at quarantining victims snuffed out the epidemic. The civet lives in the rainforests of southeast Asia and Indonesia. As the forests disappear, the animal is forced into the hands of hunters.

But the most deadly disease to cross from jungle animals to humans in recent years has of course been HIV, the virus that causes AIDS. It is thought to have jumped the species barrier from monkeys, who suffer a simian form of AIDS, around 1960. It made the jump through some sort of direct blood-to-blood contact with humans. As with Ebola, the conduit was probably infected bushmeat. And it wasn't a one-off. French researchers in Cameroon recently analysed the leftovers of meals where the meat from eight hundred monkeys was served. They found that 130 of the monkeys were infected with the simian equivalent of HIV. AIDS has already infected 42 million people, killed 25 million and reduced life expectancy by twenty years in several African countries.

In the jungle environment, viruses seem to spread from apes to humans quite often. In 2004, doctors wrote in the Lancet how they had gone looking in Cameroon for the Simian Foamy virus, which is endemic in wild primates. When the doctors tested local villagers, they found many of them carried the virus too. All the carriers had hunted and killed wild primates. Detailed DNA tests showed they had picked it up variously from mandrills, gorillas and a type of monkey called De Brazza's guenon. It didn't seem to hurt them; but, according to Nathan Wolfe of Johns Hopkins University in Maryland, it showed that 'primate viruses are infecting humans on a much more frequent basis' than anybody thought.

FRUITS OF THE FOREST

What are rainforests for? Are they there for our aesthetic pleasure, or to protect the planet's biological diversity? Are they there to provide hunting grounds for traditional rainforest tribes, or to offer 'environmental services' for the planet, soaking up carbon dioxide, keeping the rains falling, and so on? Do they have intrinsic value beyond mere human concerns, or are they about making money? Are they worth saving – and, if so, should we try to harvest the fruits of the forest in a sustainable way, or keep the forests as they are for future scientists to go prospecting for cures for diseases? Perhaps the answer is different in different places and at different times. But if so, how do we judge what should apply and when? To protect them, should someone be in charge of the rainforests – or is ownership part of the problem? One thing is for sure: saving the forests is an increasingly complex business.

BIOPROSPECTORS – OR BIOPIRATES?

For some, the rainforests are a biochemical laboratory of stupendous potential. 'At the dawn of the twenty-first century, many people believe the natural world has nothing left to offer us in the way of new medicines. This could not be further from the truth.' So says Mark Plotkin, an American ethnobiologist and acolyte of the legendary Harvard man Richard Schultes. 'Mother Nature has been creating weird and wonderful chemicals for 3 billion years, and we're only beginning to sift through these hidden treasures. While today's laboratories can synthesize new molecules at a pace unimaginable a few decades ago, nature provides the optimum starting points. Time and again, we find that plants and animals make strange molecules that chemists would never devise in their wildest dreams.'

But this is a very unusual kind of laboratory. Finding those drugs depends in large part on tapping the knowledge of the dying bands of people who live in the forests and carry the knowledge of their ancestors about traditional medicines. And, in the jungle, one man's dispassionate researcher is another man's latter-day conquistador. To some, bioprospectors are no better than biopirates.

Take the case of Conrad Gorinsky. Nice chap, a scientist, born in the South American rainforest state of Guyana, the son of a Polish father and a mother from the local Atorad tribe. He was busy in Britain through the 1990s, arguing the case for saving the rainforests by demonstrating the potential economic value of the plants they contain. He foresaw a partnership between the forest people, who know about the plants, and outside scientists, who can sift through their knowledge and broker its sale to outside companies, especially drugs manufacturers. For him, the peoples' knowledge was at least as commercially valuable as the plants themselves. Companies could spend decades prospecting to find out what the natives could tell them in an afternoon.

So Gorinsky set up an organization called the Foundation for Ethnobotany. His ideas brought him prestige in the environment community at a time when harvesting the 'fruits of the forest' was seen as the key to their survival. He was awarded a fellowship at Green College, Oxford University, which was run by a distinguished former UK ambassador and environmentalist, Sir Crispin Tickell. I interviewed Gorinsky in 1992 for the *Independent* newspaper, and found a man following in the footsteps of Spruce and Schultes.

He had a passion to collect the indigenous knowledge of the Wapishana people, who live in the border lands between Guyana and Brazil, and to help them sell their knowledge about the forest plants to the outside world. 'Governments and Western multinationals are carving up the forest dwellers, who are being offered no title to their genetic wealth,' he told me. Without people like him retrieving that knowledge before the last tribal experts die, he explained, 'the world will undergo a kind of amnesia that will be greater than any in human history. The extinction of knowledge will be even quicker than the extinction of species.'

Gorinsky's special botanical interest was the greenheart tree (*Ocotea rodiaei*), a tree with many local uses and whose timber is exported for building ships and use in marine engineering. He says he spent his own money on researching the tree, gathering local knowledge about its attributes and isolating the active ingredients of bark, leaf and nut back in the laboratory. The Wapishana people told him that they grate the tree's nut, which they call tipir, and eat it to stop haemorrhages and prevent infections. It is also effective as a contraceptive and can be taken to induce abortions. A potent drug, clearly it was essential to take it in the right dose. Tradition among the Wapishana had it that only the spirits of the dead could tame its power, which can otherwise kill. Gorinsky isolated a protein from the nut, an antipyretic that he says prevents malaria fever and can treat tumours. In his eventual patent applications in Europe and the US, he renamed it rupunine after a local river on Wapishana land.

Gorinsky also investigated a bush called cunani (*Clibadium sylvestre*). The Wapishana told him that a wad of its leaves thrown into a river kills fish, but leaves their flesh safe to eat and the water clean. The cunani leaf, he found, contained polyacetylenes. These act on the neurones of muscles and, besides killing fish, can prevent heart blockages. He renamed the active ingredient in the leaf cunaniol and again registered for a patent on its applications. 'This is a chance to demonstrate to the developing world how tropical rainforests can be saved, while helping their countries become rich,' he told me. But, while offering a share to the Indians, he also wanted to make some profit from his efforts and entered into deals with pharmaceutical companies to market the compounds.

Then things got messy. Tribal members who had previously co-operated with him said that they had used the greenheart tree and cunani bush for generations. The knowledge of the uses of these fruits of their forest was Wapishana property, they said, and not for Gorinsky to profit from, even in partnership with them. The tribe went to court and succeeded in overturning his patents. Gorinsky hit back. Apparently throwing aside the language of partnership, he told the *Guardian* newspaper: 'I made all the intellectual effort and spent thousands of dollars out of my own pocket. Would the Indians ever invest in this? The Wapishana just inherited the greenheart. They don't own it.' Gorinsky regards himself today as a bruised and broke environmentalist.

Things were no better in neighbouring Venezuela, where Gorinsky's colleagues at the Foundation for Ethnobiology had been working with three Venezuelan universities to survey indigenous plant remedies in the Orinoco basin. Venezuela, said the project director Fabian Michelangeli, ranks sixth in terms of world biodiversity, with at least 25,000 plants, of which around two thousand have medicinal uses. But as work got going the politics got nasty. A church-backed local indigenous group complained that the scientists were stealing their knowledge, and the government withdrew support for the research. 'It's so frustrating. In the year we were working we had collected 1,200 species, of which three hundred could have medicinal applications,' said Michelangeli. 'Twenty have shown powerful antiviral, antibiotic or antifungal properties, and two contain components which seem to attack breast-cancer cells. It would be such a shame if the world was deprived of a cure for AIDS, for example, because of political manoeuvring.'

* * *

Since then, researchers have gone back to the drawing-board in both Venezuela and Guyana in an effort to rebuild relations with the locals, whose knowledge of plant species in the forest is vital to the success of their work. But the story encapsulates a wider problem. Were Gorinsky and his colleagues ripping off the Indians – or seeking reasonable recompense for years of research, without which the outside world might have remained ignorant of the Wapishana's knowledge? And if we might have sympathy with Gorinsky as a lone idealist in a political jungle, what view should we take about Swiss pharmaceuticals giant Novartis, which gave a Brazilian government body 4 million dollars to be allowed to prospect for valuable chemicals among ten thousand plants in the Amazon? Or the US National Cancer Institute, which has already collected and genetically screened more than fifty thousand plant and animal samples from thirty countries in an effort to find drugs that fight cancer? Or the US Army, which holds gene patents on numerous traditional treatments for tropical diseases, including a plant called mamala from Samoa, and also on forest micro-organisms that might one day be used in biological warfare?

The secrets of rainforest species offer a vast pharmacy of potentially extremely valuable chemicals and other materials. More than a hundred of the top 150 prescription drugs in the US are derived from plants and animals, mostly from the rainforests. Dozens of major drugs companies have offices in Brazil that send out scientists searching for plants, as well as interviewing rainforest dwellers to learn their secrets. As a result of this research, ethnobotanists have, in recent years, found treatments for Hodgkin's Disease and childhood leukaemia lurking in the chemistry of the rosy periwinkle, a tiny forest-floor plant from Madagascar; a blood thinner now being marketed by Merck came from a native cure for snakebite; and an anti-viral drug that fights HIV first surfaced in the pharmacopoeia as a Samoan herbal remedy for hepatitis.

But the wisdom that is delivering these findings is being lost at a great rate. 'Tribal knowledge represents tens of thousands of years of human experience. To lose it now, just when we have developed the scientific tools to evaluate it as a source of new drugs, crops, ecological insights and conservation techniques would be extraordinarily short-sighted,' says Paul Cox, director of the US National Tropical Botanic Garden in Hawaii.

But how much do rainforest dwellers gain from this process?

Usually very little. There is truth in the charge of José Leland Barossa of the Brazilian government's native peoples' agency that the partnership process about which scientists talk is not a meeting of equals. 'The scientists congregate in some small frontier town. They ask the Indians what they would do if they had a headache, muscle pain or a bad stomach. The local people then take them into the jungle and show them which plant they would use to cure those symptoms. The scientists pay the Indians a little money, then take the plant back to their labs. There, they discover the principle by which the plant works and sell their preliminary research on to the pharmaceutical companies for development.'

The trouble is that it is very easy now for the scientists and others who profit from these findings never to return to the forests. If the natives are shareholders in the process, they never get an annual report, let alone a vote on corporate strategy. In the distant past, the American Indians managed to hang on to the secrets of cinchona for several hundred years, before botanists stole the seeds and chemists extracted its essence. Even a few decades ago, it could take some time before the transfer was completed. When Schultes made his investigations into curare in the 1940s, the drug continued to be harvested by tribal people and sold to American and European hospitals, who used it as a muscle relaxant during surgery. But, as Schultes's obituarist Herbie Girardet pointed out, 'Once Schultes and the chemists back home had unravelled its secrets, it was synthesized in laboratories and the trading relationship with the forest people ended.'

Many plants today are captured and transplanted or

synthesized almost as soon as the compounds are found. The value leaves the rainforest with lightning speed – but not always. To this day, the Brazil-nut tree grows well only in its old rainforest habitat because of the complex ecological synergies that it needs to thrive. The active ingredient in the Madagascan rosy periwinkle can be made synthetically, but it is a difficult process and the major manufacturer of the drug still buys and processes several tonnes of the flowers each year to satisfy demand. But the price is low and royalties do not reflect its real value to their patients, nor the 'intellectual property rights' of the Madagascans who first showed the drugs manufacturers where to find the flower and what they used it for. Environmental economists estimate that the commercial value of this plant runs to hundreds of millions of dollars a year.

* * *

But there is a double bind here. On the one hand, justice suggests that forest inhabitants should gain proper recompense for plants and compounds taken from the land, especially when they helped track them down. And the most straightforward way of ensuring that is if the plants are harvested and purchased directly, rather than being analysed and their compounds synthesized. On the other hand, the harvest may itself be very damaging to the forests. Like logging and hunting for meat, hunting for medicinal plants can sometimes be a death knell for forests. Markham was not wrong back in the nineteenth century when he warned that

Europe's demand for cinchona threatened the survival of the plants that provided it.

Take a look at the slopes of Mount Cameroon in Central Africa and you will see what happens when a new bio-rush takes hold. The medicinal El Dorado here is the bark of *Prunus africana*, the African cherry tree, which contains an ingredient that reverses what Africans call 'old man's disease' – the swelling of the prostate gland. The bark was the basis of an ancient African remedy; and, in the 1960s, the potion was patented by a Frenchman as pygeum. To this day, nobody has identified the active compound, so chemists cannot synthesize it. As a result, 220 million dollars' worth of the powdered bark is sold every year to ageing men in Europe and North America. But on the slopes of Mount Cameroon, local farmers spent the 1990s felling *Prunus africana* trees at eight times the rate of regrowth, in order to sell the bark to a French processing company. They saw little profit from their endeavour, but at the end of the decade the forests became so depleted that the trade crashed, the processing plant closed and the farmers were out of work. The tree takes fifteen years to regrow. With other sources also in a parlous state, it is not only the Cameroonians who have lost out. Thanks to over-harvesting, ageing men across the world 'are about to lose this remedy', says Tony Simons at the Nairobi-based International Centre for Research in Agroforestry.

Natural remedies are becoming billion-dollar businesses; and in many places they are wrecking the natural habitat of the plants in the process. Two-thirds of the fifty thousand known medicinal plants are still harvested from the wild.

Increasing numbers are being added to the international Red List of endangered species, and as many as a fifth of them may now be endangered. One of the latest to get a listing is tetu lakha, a small rainforest tree from Sri Lanka that, like *Prunus africana*, is used in Europe to fight cancers.

* * *

It is easy for rainforest dwellers and their representatives to become carried away by the potential of the rainforest pharmacy. One Reuters article filed in 2001 from the Malaysian province of Sarawak in Borneo began: 'Abang Anak Raba never went to medical school. He never even went to school. But what he knows about Sarawak's plants could be worth a fortune to Western drugs firms. Walking in hot, humid forest near his Iban longhouse, Abang stops often to slice off tree bark or pick a leaf, detailing native cures for high blood pressure, diarrhoea and childhood bed-wetting with the nonchalance of a family doctor.' This is truly astounding knowledge. But such potions are also widely available in rainforests from Brazil to Brazzaville, Ecuador to Uganda. Around the same time as Reuters reported on Abang's plants from Borneo, a paper in the journal *Science* reported that one elderly Samoan woman knew of 121 herbal remedies from ninety local flowering plants and ferns. There is just too much of a good thing out there.

There are so many remedies known to rainforest communities round the world that it is a buyer's market. Almost every African community has some local remedy for symptoms like fever, dysentery, period pains, headaches or liver conditions. Of course, some are better than others, or of more practical use to Western drug companies, but nobody knows which until the research is done. If one community won't play ball with the scientists, they will just move on down the road.

And yet the unique does sometimes show up. Searchers after nature's medicine chest still tell the story of Michael Tyler's discovery of a frog in a creek in the Australian rainforest. This was not just any frog. Tyler, from the University of Adelaide, noticed that the female swallowed her eggs, incubated them in her stomach and gave birth to tadpoles through her mouth. Why, he wondered, didn't the frog's stomach acids destroy the eggs? It turned out that she could 'turn off' her stomach acids while carrying eggs. Maybe, he reasoned, the frog produced some compound that could do that. If chemists knew what it was, then maybe they could develop a better means of alleviating stomach ailments, and acid indigestion could be a thing of the past. Maybe they could have developed such a drug, but we will never know. The frog became extinct before anyone had a chance to find out what went on in its stomach.

RAINFORESTS: WHO NEEDS THEM?

The new orthodoxy about rainforests has taken it as read that they are worth more alive than dead. Worth more, not just to science, to the planet's biodiversity and to our profound sense of wonder about nature – but more also as cash in hand for the inhabitants of the forests. It is one of the dogmas of the day that good conservation and good economics automatically go hand in hand, and that people who destroy their forests are doing themselves a bad turn, as well as the rest of us. But is it true? And what would it mean if it were not true? What if the world as a whole had a great amount to gain from protecting the forests, but that, in terms of cash at least, the inhabitants of the forests did not?

Little-noticed research published in 2000 suggested just this: that the new orthodoxy could be far from true in many rainforests. The romance of the rainforest, says Ricardo Godoy, an anthropologist from Brandeis University in Waltham, Massachusetts, has seduced scientists and the wider world into inflated notions about the true economic value of the forests to their inhabitants. Godoy reached this conclusion after conducting the first detailed household inventory of the fruits of a rainforest. He sent teams of students into a remote region of eastern Honduras in Central America, to catalogue what Indian villagers harvest from their forests and to ask what that harvest fetches at market.

For more than a decade, biologists have taken it as an article of faith that most rainforests are a rich source of food, medicines and traditional building and craft materials.

We have seen that many forests contain lots of medicinal plants that are still widely used, especially in remote regions away from doctors and clinics. And any market near a rainforest will show you that foods, too, are widely harvested. But what are these products worth?

In the most famous study, Charles Peters of the New York Botanical Gardens found that almost every part of the Amazon rainforest in Peru contained hundreds of valuable species. In 1989, he valued the annual harvest at 650 dollars per hectare – more than twice its value as either timber plantation or cattle pasture. The conclusion was obvious, and a great relief to biologists and conservationists. The forests were worth more to locals intact than if they were cut down. As Peters put it: 'Without question, the sustainable exploitation of non-wood forest resources represents the most immediate and profitable method for integrating the use and conservation of Amazonian forests.' The findings have been widely cited ever since. Among other things, they gave a huge push to the idea of creating 'extractive reserves' – areas of rainforest protected from logging, and dedicated exclusively to the harvesting of nuts, fruits, wild rubber, plant medicines and other natural products.

But the suspicion has grown that some ecologists have been guilty of wishful thinking. As Godoy points out, few of them undertook actual inventories of harvested products. The researchers, he says, 'focused on what they saw as the potential value of the forest, rather than what was actually being taken from the forest'. That sounds like bad economics as well as fanciful thinking. A small amount of valuable rainforest fruit might fetch a high market price.

But harvesting ten times as much will more likely just flood the market and send prices tumbling. Peters, say his critics, did not address this. Nor did he ask how the harvesters in remote jungles would get their products to market, even supposing that there was a market.

Godoy's own two-year study painted a very different picture of the fecundity of the forests. He concentrated on a detailed analysis of the domestic economics of thirty-two households in two villages, Krausirpe and Yapuwas, both deep in the Honduran jungle in the heart of the Tawahka Anthropological Reserve, and home of the Tawahka Indians. 'Ours is the first attempt to physically measure the goods that came into the house, whether plants or animals, and to put a value on each,' he says. There were many products, including fruit, fish, wild game, medicinal plants, firewood and construction materials. Not all of these products had a local cash value because not all were traded in the village. In these cases, Godoy asked villagers how much of a commercial product – salt, for instance – they would be willing to give up in exchange.

The result was a shock. Godoy's team found that the annual harvest from a typical hectare of rainforest around the two villages was worth a measly twenty dollars, roughly a twentieth of the value found in Peters' Amazon study. This is rough news for environmentalists, but hardly surprising, according to Godoy. 'People in the rainforest are poor. If the forest produced high economic value to these people, they would not be poor.' Honduras is the second poorest country in the western hemisphere, and the Tawahka Indians among its poorest, most isolated, inhabitants. What, for biologists, seems like a green heaven is, for them, a green slum.

Many environmentalists and even economists blame corruption, international debt, urban poverty, globalization and the international timber trade for the destruction of the rainforests. They argue that if the forest people had full control of their land and the right to manage it without the risk of take-over by land barons and corporations, they would protect it and harvest its riches. Not so, Godoy says. Rural people chop down the forests 'because they are poor and stuck with a nearly worthless asset, not because they lack security of tenure,' his economic analysis suggests.

For such reasons, Hondurans have removed half the country's forests in the past forty years. Godoy's study suggests that this makes reasonable economic sense for the people on whose land the forests sit. If people want cash, investing in a chainsaw is the best way to get it. Can a fifth cavalry of ecologically minded capitalists change this gloomy prognosis? Some have certainly tried to develop international markets for forest products. The Body Shop in Britain and Ben and Jerry's ice-cream in the US are high-profile initiatives of the past decade. And forest campaigners have worked to create extractive reserves and encourage the hope that 'bioprospectors' might save the day by buying the rights to comb forests for plants that might cure cancer or provide some other valuable chemical for pharmacologists, food manufacturers or industrialists.

But after more than a decade of effort, most extractive reserves remain poverty traps and those bioprospectors have seldom left behind much of the promised wealth. David Simpson of the Washington think-tank Resources for the Future concludes: 'When it comes to commercial prospecting in natural resources for new products, the value of biodiversity is not as high as some conservationists suppose.' The truth is that, for all the effort, the economics of rainforest destruction have barely altered.

Godoy is not saying that the forests should be allowed to disappear. Far from it. It goes without saying that the forests have an intrinsic value that cannot be measured in dollars. But even if we restrict ourselves to more prosaic issues, there are practical reasons why we need them – like sustaining the planet as a place fit for life. The rainforests perform vital global functions, capturing carbon dioxide from the atmosphere, stabilizing the climate and harbouring biodiversity. And, for individual nations, the forests regulate rainfall, protect soils, store water and reduce flooding. In 1998, two years before Godoy's survey, massive floods in Honduras killed some ten thousand people during Hurricane Mitch. The floods were probably made much worse by the country's deforestation. Steep hillsides that had lost their forests were unable to soak up the rain, and succumbed to landslips. Recent efforts at putting a dollar-value on such 'ecological services' to the planet have valued the rainforests at between 1,700 and 25,000 dollars per hectare. Even the lower figure is almost a hundred times the paltry harvest of the Honduran villagers in Godoy's study.

Godoy's message is that 'tropical rainforests are worth more for their global than for their local value'. And that means that if the world wants to save the rainforests, as it should, then it will have to pay the forest inhabitants well to protect them. 'It might be politically unpopular to argue for an outright subsidy,' says Godoy, 'but I don't see any other way out of the impasse.' While we continue to hold romantic ideas about the dollar-value of rainforests for their inhabitants, we will continue to demonize those among them who seek to better themselves by chopping them down. We have to get real: the planet needs the rainforests, and we, spiritually, need them. We should find a way of paying for their preservation.

What do they make of all this in the Tawahka reserve? Probably not much. Just before Godoy's survey, the Honduran government took away many of their rights in the name of the environment. It declared the whole region a Biosphere Reserve and gave control over it to the state Forest Development Corporation. The area had 'great economic potential as far as biodiversity is concerned', it declared. So whatever wealth is there, the Tawahkans are unlikely to be getting their hands on it. Fruits of the forest? You can almost see them canoeing into town to buy chainsaws.

ON PATROL WITH THE GREEN CORPS

When the forests are on a one-way trip to destruction, perhaps almost any methods are justified. The Chinko river in the Central African Republic is known as the 'river of elephants'. It also has huge herds of hippos and crocodiles, buffalo and giraffes. But poachers now invade every year from Sudan, sweeping across the savannah grasslands into the jungle. By some estimates, 95 per cent of the animals have already been lost, while ivory heads back across the border, destined eventually for the Far East, and smoked bushmeat is carried down jungle paths across Central Africa and even on to European tables. But, for some, Bruce Hayse's efforts to save the forests smack too much of the brutal fiefdom in Joseph Conrad's *Heart of Darkness.* Mr Kurtz goes green, perhaps.

Hayse is a doctor from Wyoming. He was a founder member of Earth First, an environment group formed in the US in the 1970s with a radical agenda of civil disobedience. After watching the death of wildlife in the Chinko valley during whitewater rafting trips, he decided to take his strong-arm tactics to Africa. He has recruited a mercenary army carrying AK-47s to police the jungles of the Republic, where official policing in national parks has virtually disintegrated.

In 2001, President Ange-Felix gave authority over the country's Chinko river basin – an area of almost a quarter of a million square kilometres – to Hayse and his force of local villagers and foreign mercenaries, who have been put under the immediate control of a former South African commando whose true identity only Hayse knows.

'The goal is not to kill people,' Hayse told the Press Association in a rare interview. 'But you can't just declare a national park and assume that the animals will be safe. There will be some confrontations, and you have to assume there will be gunfire.' He says his forces are no more heavily armed than the poachers themselves; but poachers have certainly been killed, and he takes no prisoners. Jailing is not an option, he says, because 'you have to have a jail first'.

Regular environment groups disapprove. WWF, which won its spurs forty years ago protecting African wildlife in the aftermath of independence from colonial rule, is trying to disengage from direct wildlife management and give control of parks back to the national governments. It once paid for the Kenyan Wildlife Service to buy helicopter gunships to shoot down elephant poachers, but now takes a very different view. It has had enough of shooting people to save wildlife. 'Allowing a private militia run by expatriates to control the situation using lethal force against Africans will backfire on the government and hurt conservation in the region,' it says. But militant conservationists say there is no alternative.

* * *

For people such as Hayse, wildlife hotspots like the Chinko valley have to be saved at almost any cost. Some contain large numbers of big game; others are simply ancient centres of biological diversity. If we have our history right, many may be the ancient rainforest refugia that survived the ice ages. The argument is this. Most of the world's species, and virtually all of those living in rainforests, can be found in relatively small patches of forest, areas of super-biodiversity, or hotspots. Norman Myers, an independent British conservation scientist who popularized the idea, reckons that half the world's land species can be found in twenty-five hotspots that, in total, comprise an area no larger than that of Greenland. He reckons that this 1.4 per cent of the land surface of the globe is home to 44 per cent of all the plants and 35 per cent of the land animals.

Most of these hotspots are in rainforest regions such as the western Amazon, the forests of West Africa, Madagascar and the islands of southeast Asia. Save them and you will save much of the world's biodiversity, Myers argues. But not many take Hayse's view that they should be defended with quasi-military private forces. A great many environmentalists place their faith in national governments, sometimes responding to international environmental pressure, to declare these great rainforest hotspots national parks and defend them with the power of law. Others take the view that the dollar might do the job better. And the idea of concentrating conservation efforts on these hotspots has been formally adopted by at least one large US-based private charity, Conservation International, which buys up wilderness around the world. The charity could, conceivably, one day become biodiversity's landlord.

These conventional approaches all broadly try and separate rainforest inhabitants from the rainforests. They do it more or less liberally. Forcing people to leave their villages is generally frowned on by the international environmental community. Forcing them to give up hunting or slash-and-burning in the jungle is regarded as, at best, a last resort.

One of the totems of liberal-minded park conservation for some years has been the Korup National Park in south-west Cameroon. It tries to combine the economic interests of locals with hard-nosed conservation. The forest is a hotspot. It contains more than three thousand known species of plants and animals, including leopards and a quarter of all the African monkey species. To protect them, the Cameroon government established the Korup park in 1986 with the help of the environment group WWF, which has been deeply involved in its management. Once, the park contained six villages, home to over a thousand people; and there are still some sixty thousand people settled nearby. Traditionally villagers have depended for much of their protein on hunting in the forest; but, today, hunting is endangering the survival of species like the rare red colobus and drill monkeys. The park authorities have expelled the villagers and banned hunting in the park (in theory, if not always in practice), as well as the gathering of fruits eaten by the monkeys and elephants. By way of compensation, it is investing in farms in a buffer zone around the forest, where it has encouraged villagers to grow traditional crops, such as cassava and manioc, as well as crops new to the area, like beans and yams.

Most controversially, the park authorities are also building the first all-weather roads into the zone around the park. The idea is to allow the villagers to sell their crops to outside markets more easily than before, so raising the incomes of villagers and reducing their dependence on the fruits of the forest. The Korup project, say its managers, 'can be a model for our neighbours and all those who want to preserve their rainforests while improving their living standards'. The danger, from a traditional conservationist viewpoint, is that the roads will bring in migrants and poachers and provide easy access for logging companies. The jury is still out. It remains a test case.

* * *

But many critics of the orthodoxy exemplified by Korup see efforts to separate people from the land as frequently counter-productive and ultimately doomed. As many as a billion people live in these hotspots, many of them among the world's poorest people, and they desperately need the land to hunt or to make a living through farming. 'Endangered species, essential farmlands and desperately poor humans often occupy the same ground. We cannot separate people and wildlife. There is no way of conserving biodiversity that way. We have to find a productive balance with nature,' says Jeff McNeely, chief scientist at the Swiss-based World Conservation Union.

McNeely says that we need to find better ways to live with nature, through more ecological farming: not through separation, but through coexistence. And the discovery that mankind has done this before – that most of our rainforests are far from being virgin wildernesses and already represent compromises between people and nature – should encourage us in that direction. There is, says McNeely, great potential to develop more ecological methods of farming that will not destroy the forest wholesale, but which will allow an accommodation between farmer and forest, between conservation and economic development. Three stories – one from Bolivia, one from Cameroon and one from the Amazon – illustrate in their different ways what may be possible, but also the barriers along the way.

* * *

We start with Andreas Chileno. He dreamed of a new start when he left his smallholding in Cochabamba, an overcrowded farming community in the Andean mountains of Bolivia, for the forested lowlands of Santa Cruz in the east of the country. This was back in the 1960s, when the Bolivian government was encouraging migrants by offering free plots of forest to clear and farm. Millions came, both from the poor highlands and the slums of the cities. But Chileno found that whenever he cleared the jungle from part of his plot, the rain quickly washed the fertility from the soil. Within a couple of years it was full of weeds and he had to abandon it and clear another patch. With 50 hectares in all, he didn't need to farm his entire plot all the time, so he became a kind of shifting cultivator in the forest, moving on to another part of his allocation every few years.

'We tried,' he says. 'We worked the land, bit by bit cutting down my forest. But it rained and rained and rained. The mosquitoes were insufferable. We experienced terrible suffering.' On his old upland farm, he was used to planting maize and wheat, but here he had to grow rice and cassava. 'At the beginning the rice was wonderful, but from then on it never produced the same. Now the only thing this land is good for is grass and livestock.'

His problems are typical of those faced by the many millions of people who have moved into the Amazon rainforest in the past forty years in Bolivia, Peru, Ecuador, Colombia and, of course, Brazil. Life has been hard. As each piece of land is exhausted, they have moved on; but eventually they have to return to their original cleared area, usually long before the land has had time to recover. Each time the farmers return, the soils are worse, the weeds grow higher and the crops grow less well. With little cash for fertilizers and other inputs, the farmers are clearing ever more forest, leaving behind a trail of eroded soils, degraded vegetation and broken dreams.

Shifting cultivation can be sustainable in the forest, and slash-and-burn can work. It has a long and successful history in lightly populated rainforest areas, where farmers know how to improve their soils. The evidence, as we have seen, is there in the wonderfully fertile black earth across the Amazon basin. But when migrant farmers move into the rainforest, things can be very different. They lack the skills to farm the forest successfully, and very often there are too many of them to allow the forest time to recover.

But Chileno and his fellow farmers in the Santa Cruz forest are working with British and Bolivian agronomists to turn things round. Their project is called 'alternatives to slash-and-burn'; and, if they can crack it, then the lessons they learn could have huge applicability across the Amazon and in other rainforest regions of Africa and Asia. The aim of their project is to stop the shifting element in their cultivation, to allow the farmers to settle on fixed plots rather than hacking out ever more forest. To do that, they will have to find ways of ensuring that those plots become more fertile rather than turning to useless scrub. They need to replace the vicious cycle of declining fertility with a virtuous cycle of improved farm yields and improved soils. That way the remaining forests will be protected and the farmers' incomes will improve. The chance of a return to a settled life would bring other benefits for farmers, with access to better healthcare and schools for their children.

One of the young agronomists working on the project is Emilio Chileno, son of farmer Andreas. His work, he says, is a new kind of agricultural research based on the evidence of real farmers' fields rather than research stations. 'Farmers like my father formed a very important part of the project team,' he says. The researchers didn't come with a blueprint. They came to learn as much as the farmers. They tried many systems for improving soils. They introduced cover crops and fruit trees to reduce erosion and stop the growth of weeds. They planted leguminous crops, which put nitrogen back into the soil, alternating rice and leguminous beans to improve soil fertility. Above all, the researchers encouraged farmers to diversify and try out new things. Different solutions worked for different farmers on different farms. But the variety is important. Fruit trees such as citrus, pineapple and bananas, and timber trees such as mahogany, promise long-term income. But other crops and even livestock are needed, especially to provide income while the trees are still growing. By the end of the project, four-fifths of farmers were adopting new crops, and most said they would continue. It is early days, but it shows there is hope.

* * *

A second approach, perhaps the most audacious, is to go farming in the rainforest itself, under the shelter of its canopy. In the forests of central Cameroon, on the edge of the jungle proper, one-time environmental pariahs are being recast as saviours by environmentalists. Joseph Essissima has seen it all. He first planted his cocoa trees in the bush outside Yaoundé, the capital of Cameroon, sixty years ago. He hacked down the jungle to fill the confectionery shelves of Europe and North America, and, ever since, ecologists branded him and his fellow cocoa farmers as environmental pariahs. But now those same ecologists have come back to take a second look – and now they are praising him. Even more remarkably, they want to help him make more money out of his trees so that he can plant some more. They say that planting cocoa could help save Africa's greatest surviving rainforest, which stretches southwards from Cameroon across the Congo basin. So who has changed? Joseph or the greens?

'Cocoa has been an important agent of deforestation during the twentieth century,' says Francois Ruf of France's Centre for International Co-operation in Agronomic Research and

Development in Paris. For a century or more, clearing forests to plant cocoa has helped deforest West Africa, with Ghana and Côte d'Ivoire at the forefront. Reports in 2001 that cocoa plantations were being worked by child labour, the so-called 'chocolate slaves', only enhanced the industry's unsavoury reputation. But, in Cameroon, cocoa farms employ no slaves and generally do not clear forests to plant cocoa. They do their planting within the forest, beneath its canopy, making the forests more valuable and more worth protecting, says Ruf. Here, at least, the forest has real worth. Most Cameroon cocoa is grown by peasant farmers on smallholdings of a hectare or less, close to or actually inside the forests. Cocoa cultivation here is unusually benign to the environment. If that approach could be extended, then, says Ruf, 'in the twenty-first century, cocoa may switch from being an agent of deforestation to an agent for reforestation'.

Joseph's cocoa garden feels more like a rainforest than a farm: dark, dank and full of life. The cocoa bushes are, in effect, an under-storey beneath the forest canopy. There are cocoa trees, but also many others, some indigenous and some planted. Various fruit trees are dotted around: orange and mango, avocado and cherry. According to Joseph, some original rainforest trees have been kept for their timber, for medicinal bark, and so their canopy can provide additional shade. Of one tree he says: 'We keep this one because it attracts caterpillars that we eat.'

These cocoa forests in Cameroon are quite unlike the monoculture plantations of Côte d'Ivoire, the continent's biggest cocoa producer. They are biologically very diverse, with more than half as many species as a natural forest, says Jim Gockowski of the International Institute of Tropical Agriculture (IITA) in Yaoundé, who has studied the methods of the local farmers. What is more, he says, Joseph's smallholding is as fertile as when he first planted it. The number of earthworms – a key test of the forest floor's ability to recycle nutrients – is almost as high as in a natural rainforest. 'By maintaining a shady canopy of diverse forest species, these farmers manage one of the most biologically diverse land-use systems in Africa,' says Gockowski. It may not be virgin rainforest, but if the farmers didn't plant cocoa, the country's fast-growing population would be clearing the forest wholesale and planting maize or oil-palm or turning it over to cattle. Good news, surely.

Across southern Cameroon, large areas of former rainforest land now lie fallow after the exhaustion of their soils by farming. Yet, in the midst of this exhausted landscape, the cocoa forests that were once dismissed as just another scar on the natural landscape are green oases. 'In many ways, the environmental benefits of a closed, natural forest are now being provided by cultivated forests of cocoa and fruit trees,' says the IITA's station chief in Yaoundé, Stephan Weise. So, he asks, 'why not convert the large areas of unused former forest into cocoa forest?'

A few farmers are taking up the challenge, such as Madame Abome, whom I met not far from Joseph's smallholding, growing cocoa trees and bananas in abandoned maize fields. But there is a big problem: the international price for cocoa. Most farmers have been caught by a downward spiral in the global price; so, rather than planting cocoa beneath the rainforest canopy or on abandoned former forest land they are instead tearing up cocoa in favour of arable farming. For many years, cocoa was a profitable crop in Cameroon and the government guaranteed good prices. But in the past decade, a combination of the privatization of the country's marketing system and a collapse in the international price of cocoa has impoverished cocoa farmers. 'The government used to be like a father to us,' says Mani Alexandre, another of Joseph's neighbours. 'Now buyers can pay what they like.' A global free-market in cocoa is wrecking the Cameroon forests.

According to Gockowski, the fall in cocoa prices has 'led directly to a very significant increase in forest clearing' in Cameroon in the 1990s. Thousands of farmers left their cocoa forests to decay and cleared forest to plant maize, groundnuts or oil-palm. Today, they call cocoa an old man's crop. 'Young people don't have cocoa production in their heads. Prices are too low,' says Mani. The tragedy is that at the very time when cocoa has emerged as an environmentally friendly crop, its profitability has slumped. Soon, maybe the chocolate slave-masters will be in business here.

Enter the chocolate companies. Firms like Mars and Cadbury's buy cocoa through the big trading and milling conglomerates. Martin Gilmour, UK-based cocoa research manager for Mars, says: 'We would like to see farmers get higher prices for their cocoa. It would be better for both of us.' Like the farmers, Mars claims an interest in the ethical and environmentally sustainable production of the crop. 'We find it interesting that they appear to grow cocoa in Cameroon in a more sustainable way. Their agro-forests are ecologically speaking almost as good as natural mature forest,' says the man from Mars. The companies are funding research aimed at improving cocoa and fruit-tree varieties and fighting the diseases that are running rampant through the ill-tended cocoa forests. But it will be useless, unless the farmers can win a proper price for their product.

Many ecologists believe that agro-forestry on the blueprint of the Cameroon cocoa forests is the only way that many of the world's rainforests can be saved. And yet there is a real risk that the blueprint itself will be destroyed before it can be properly researched and copied. As I left the forest, one young boy – the son of a cocoa farmer – came up to me and asked simply: 'What does chocolate taste like?' His family, he said, could not afford the price of a bar.

* * *

Another way of making a living in the Amazon may lie with the estimated half a million descendants of the old rubber tappers who still live in the forests and small towns of the western Amazon. An estimated seventy thousand individuals still tap wild rubber and harvest Brazil nuts, palm hearts, tonka beans and other 'fruits of the forest'. Many also eke out a living by fishing, hunting and small-scale agriculture. A century ago, the tappers were seen as vanguards of rainforest destruction, but today, in a turnaround rather similar to that of the cocoa farmers of Cameroon, they are now seen as one of the forest's defenders.

If you have heard of the rubber tappers of the Amazon, it is probably because of one man, who sprang into the headlines in the 1980s at the height of international concern about the fate of the Amazon forests. His name was Chico Mendes, leader of the rubber-tappers' union in the Brazilian state of Acre. He was a descendant of a tapper who went to Acre at the time of the brief rubber revival there during the Second World War. He was befriended by Mary Allegretti, a young Brazilian anthropologist who had written a thesis on how the Acre rubber tappers were making a living in the forest without destroying it. While all around them cattle ranchers were buying up old rubber estates and reducing them to infertile pasture, Mendes and his fellows were harvesting the forest's wealth.

Mendes had come to Allegretti's attention when he ran unsuccessfully as a Workers' Party candidate in state elections in 1986 on a programme of stopping the cattle ranchers from invading the land of the tappers. Allegretti told his story to Western environmentalists, who saw a way to promote a sustainable future for the Amazon in the face of its wholesale destruction. Mendes became briefly more famous outside Brazil than within as he was whisked off to meetings with environmentalists in Miami, bankers in New York and senators in Washington.

Mendes told his audiences that all the development grants being pumped into the Amazon were destroying rather than developing the region. Instead, he proposed that large areas of rainforest be given over to 'extractive reserves', where locals would be free to form co-operatives that could profit from rubber latex and other fruits of the forest, but where nobody would be allowed to cut down the trees. It was not clear where his socialist agenda ended and his environmental agenda began: Mendes was as interested in getting rid of the local rubber barons, descendants of the original boom barons of a century before, as he was in keeping out the cattle ranchers. But he was fêted nonetheless: here was a fantastic story of a man who could save the rainforest. And, in mid-1988, the Brazilian government – under pressure from foreign governments – gave way and declared the creation of a series of extractive reserves, funded by the World Bank and others. The largest, the Chico Mendes Extractive Reserve, covered half a million hectares – roughly the size of Wales.

It was a brief fame, however. Increasingly jumpy cattle ranchers sought to turn off the international spotlight on their activities and to forestall the creation of more extractive reserves. They did it in the most brutal way imaginable, by shooting Mendes dead, on his doorstep, in his hometown of Xapuri just before Christmas 1988. There was a memorial service in Washington, funded by Greenpeace and other environment groups. It turned out to be effectively the moment the world lost interest in the Amazon.

But Mendes's ghost lingers. You can see it in the extractive reserve that bears his name, and in a handful of others that today occupy some 2 million hectares of forest. They don't make the profits he had hoped, but they are still there – and so are the forests that they enclose. And, more interestingly perhaps, the spirit of Chico Mendes lives too in the political emergence of his close associates from the heady days. First came Jorge Viana, who was elected governor of Acre state in 1998 on a platform of making the state 'the Finland of the Amazon'. Trees, he said, 'are our biggest resource, our vocation and patrimony, not an impediment to modernization, and we need to learn how to exploit the forest without destroying it'.

And then there was Marina Silva, who, in 2002, became a national figure when she was appointed Environment Minister in Brasilia, in the Workers' Party government of President Luis Inacio Lula. Born to a family of rubber tappers who knew Mendes well, Silva was an illiterate maid until she came to town to find a treatment for malaria. There she met Mendes, in his pomp; inspired by him, she went to college and spent her spare time helping Mendes organize demonstrations against deforestation and the expulsion of the rubber tappers. She was elected a senator for the first time in 1994 before moving to Brasilia and national politics. 'I'm in politics because of Chico,' she says. 'When Chico was alive, to talk of defending the forest, the Indians and the rubber tappers was seen as being against progress. Fourteen years later, no one would have the courage to say such things.' And with the successors to the green hero and social pioneer Chico Mendes in power, there are no more excuses.

OIL IN THE FOREST

Could the discovery of oil wealth in the jungle help conserve rainforests? The idea of encouraging drilling rigs and mineral prospectors deep into the forest seems an unlikely way to save these cathedrals of nature. And yet, in the modern world, it could just sometimes work. Or so say some otherwise sane and sober researchers with no obvious axe to grind. They claim that these environmental pariahs can turn the tide of deforestation in many poor, forested countries, not by directly conserving trees, but by altering the economics of their countries.

The analysis comes from the Center for International Forestry Research (CIFOR), an Indonesian-based international research outfit previously mentioned for revealing cattle farming as the new enemy of the Amazon forests. Its researchers argue that mining and oil drilling can be beneficial because they provide an alternative source of income to the most destructive activities in the forests: logging and unsustainable farming. This will come as a shock to environmentalists campaigning against oil rigs and pipelines from the Ecuadorian Amazon, through Gabon in Central Africa to Papua New Guinea in the far East Indies. But CIFOR economist Sven Wunder says recent history bears him out, arguing that long-term oil booms in Gabon and Venezuela have 'wiped out agriculture and resulted in many abandoned areas growing back as forests'.

Take Gabon. Since the 1970s, this country has become one of the richer nations in Africa, thanks in large part to oil fields deep in its rainforest region. And, unlike its neighbours, Gabon has seen an increase in forest cover since 1970. Wunder says the two are linked: whereas the inhabitants of those oil-poor neighbouring nations have had to clear forests to survive, the Gabonese have made their money in other ways. Oil, he says, provides city dwellers with money to buy food rather than grow it. It encourages farmers to leave the land and head for jobs in the cities. And finally – and maybe most importantly – the money generated by the oil industry improves currency exchange rates for the host nation, making forestry and cash cropping less profitable.

Wunder says that other rainforest countries like Papua New Guinea in the western Pacific and Gabon's neighbour Cameroon would have lost more forest if they had not been able to exploit oil and mineral reserves. A functioning industrial sector and high exchange rates discourage the short-term exploitation of a country's biological wealth. Gabon's success in countering the forces of deforestation, he says, is the flipside of what has happened in Indonesia since the late 1990s, where the collapse of the currency destroyed industry, raised the value of natural resources like forest timber and led to an escalation of deforestation. Wunder admits that deforestation resumed in Gabon in the 1990s, but argues that this was because oil prices fell sharply and the pendulum swung back towards exploiting the forests.

This is a dangerous argument, of course; and Wunder agrees that his theory does not work in all cases. In Ecuador, for instance, deforestation accelerated after the discovery of oil in the rainforest because the government used oil revenues to build roads and subsidize cattle ranches in the forests. His critics say he is wilfully ignoring the local impacts of oil on forest ecosystems. And he seems to discount the effect on local rainforest-dwelling communities, whose lives may be turned upside down by the foreign invaders.

These critics point to the history of the world's second largest oil company, ChevronTexaco, which is engaged in a protracted legal struggle in Ecuador over its role in polluting the Amazon jungle in the east of the country during thirty years of exploitation of oil reserves. The main charge is that it disposed of oil waste in sixty huge, open, unlined pits that are now damaging water supplies. The court hearings began in October 2003 in the tiny town of Lago Agrio, close to the Colombian border.

Wunder agrees that oil companies need to be held to account for their local impact in the forest. 'We in no way excuse companies using environmentally destructive mining practices,' he says. Holding companies to account may be harder to arrange in practice, but Wunder is at any rate right that it is dangerous for environmentalists to ignore the 'macro-economic' effects of capital-rich industries in poor countries.

AMAZON DREAMS

All sorts of unexpected people turn up deep in the jungles. America's on-off love affair with the Amazon survived the end of the rubber boom and persisted into the twentieth century when historian James Bryce dreamed of Uncle Sam taking charge in Brazil: 'How men from the Mississippi would make things hum along the Amazon,' he wrote in 1912. 'Steamers would ply upon rivers, railways would thread through the recesses of the forest, and the already vast domain would almost inevitably be enlarged at the expense of weaker neighbours until it reached the foot of the Andes.' In place of southern slaves there were dreams of sending 'a million Chinese' into the forest to revive the rubber business and compete with the British and Dutch plantations in the Far East.

Around this time Henry Ford founded Fordlandia, when he took over a million-hectare slab of forest along the banks of the River Tapajos with the aim of establishing rubber plantations. In the late 1920s, he built a town complete with schools and hospital to house workers who would supply rubber for his Detroit car plants. But the soil proved poor and disease struck the plantations, even after he brought in supposedly disease-resistant grafts. Local Brazilians hated working or eating in the American way: there were strikes and mass firings and, in 1945, Ford sold up.

Some foreigners fared better. There is a large and successful community of Japanese in the Amazon south of Belem, for instance. They, too, showed up in the 1920s and formed a colony near the small town of Tome-Acu as part of a deal with the Japanese government, which was then plagued with unemployment. Things started badly: they tried rice paddy and cocoa, but both failed. Yellow fever and malaria carried many of the immigrants off and, after a decade, only a third of the 350 settler families remained. But they had more sticking power than Henry Ford, and gradually they found new ways to farm, intercropping pepper, passion fruit and coffee between rows of rice, cotton and beans. They used dung and mulch to improve the soils – much as the ancient Indian rainforest inhabitants once did. And they ran a co-operative marketing system that, though very different from Brazilian ways, still survives and is highly effective. Researchers say the key to their success is treating the land with care, as they would have done in Japan, where it is scarce, rather than with abandon, as most Brazilian farmers tend to do.

Even in recent times, the Amazon has remained a land where dreams are conjured and shattered. In 1967, Daniel Ludwig, an American shipping magnate, set out to build a huge industrial zone dedicated to wood-pulp production on the Jari river, along the Amazon, west of the giant estuary island of Marajo. He constructed townships for thousands of people, laid 4,500 kilometres of roads, and drew up plans for paddy fields, an 80-kilometre railroad and a power station burning timber culled from the rainforest. A pulp mill was brought on barges by sea from Japan round the Cape of Good Hope and up the Amazon. And finally he started tearing down the jungle and replacing it with his chosen tree, the fast-growing *Gmelina arborea* tree from Asia.

This was like an entire civilization in the rainforest. Some compared it with the ancient jungle societies whose archaeological remains were then being dug up by Anna Roosevelt, close by on Marajo Island. But Ludwig proved far less successful than the ancients. Bulldozers brought in to remove the original rainforest made such a mess of the thin soil that the new trees wouldn't grow. Substitute trees from the Caribbean scarcely fared any better. All told he spent three-quarters of a billion dollars over fourteen years before finally selling out for a fraction of the cost of the investment.

And more recently still, we have the Reverend Sun Myung Moon. The Korean reverend, founder of the Unification Church, or more popularly the 'Moonies', set out in the late 1990s to establish a large settlement for his disciples in the state of Mato Grosso, on the edge of the jungle. He bought 570 square kilometres of land to fulfil his dream of creating a 'kingdom of heaven on Earth, a new Garden of Eden' for his followers.

Not much, it seems, has changed since the first conquistadors headed for the Amazon almost five hundred years ago. The place remains for many the quintessential land of dreams.

APES, ANCESTORS AND A NEW HUMAN DIMENSION

For millions of years we shared the jungles with our nearest ancestors – the great apes. Charles Darwin caused outrage when he suggested that we were descended from the apes; but today, as we learn more about them, we discover how unnervingly like us they are. From sexual shenanigans to murderous mayhem, we humans do it their way, and they do it ours. And as we, the naked apes, think afresh about ourselves, so we are thinking afresh about the rainforests from whence we came. Scientists are returning to the forests to discover how to make new drugs, to find new crops and to solve the riddles of evolution and the workings of our planet. Are we returning home?

OF APES AND HUMANS

When we look into the faces of forest apes, we see ourselves. And the more we look, the more like us they seem. It is not surprising: after all, we go back a long way. For a long time we shared the forests, says Caroline Tutin, who has carried out long-term research in the Lope reserve in the Central African state of Gabon and charts what she believes is 'sixty thousand years of coexistence' between chimps, gorillas and humans. And while sharing the same jungles, we seem to have shared many of the same characteristics. Whenever scientists try to draw a line between humans and our nearest relatives, the apes later turn out to be on our side of the divide, confounding our views about both them and ourselves.

Once, scientists defined humans by their ability to find and make tools. Now, it doesn't look as if we are so special in that department. The idea began to fall apart one day back in the 1960s, when pioneering primate researcher Jane Goodall watched a chimp in the wild pick a blade of grass, trim the edges, stick the grass into a termite mound, leave it there for a moment and then pull it out. The blade of grass was swarming with termites, which the chimp ate. The chimp had, she pointed out, made a tool – a 'fishing rod' to catch termites.

It was the first report of chimps making and using tools in the wild. The German animal psychologist Wolfgang Kohler had previously tracked them doing similar things in captivity, but making animals do tricks in captivity is different. What Goodall was watching was wild behaviour, and it raised the intriguing question: if tool-making is only a human trait, then are chimps as good as human? Goodall had opened a new debate, not only about what it means to be human, but also what it means to be a chimpanzee. It certainly helped establish these animals in the public mind as our closest living relative – chimpanzee tea parties and all.

Today, there are many examples in the research literature of highly sophisticated tool-finding and tool-using behaviour among primates. Take the capuchin monkeys in the forests of Venezuela. These are the New World equivalent of the chimpanzees of Africa and the orang-utans of the East. Like their counterparts, they are clever tool makers. They are good, for instance, at selecting rocks from which to fashion an anvil on which they will break open nuts. They trawl the forest, collecting rocks that they carry home above their heads. But they are not just crude engineers; they are pharmacists, too.

Capuchin monkeys have come up with their own natural mosquito repellent as good as anything you could buy in the local pharmacy. You can see the monkeys high up in the branches near the Orinoco river, rubbing their repellent on to every part of their body, from head to foot. What the capuchins have discovered is that a millipede that lives in tree bark and termite mounds wards off mosquitoes. In fact the bodies of the millipedes contain large amounts of two powerful insect-repelling chemicals called benzoquinones. While not analytical chemists, the monkeys are smart enough to make a habit of poking around in the bark of the right tree, finding the millipedes and rubbing them on. They do this every evening when mosquitoes are at their most aggressive, and most assiduously during the rainy season, when mosquitoes are at their most profuse – so they don't seem to be in any doubt about why they are doing it.

'We think this is the clearest case yet of an animal using organic material for medicinal purposes. The chemical analysis leaves very little room for doubt,' says Ximena Valderrama of Columbia University in New York. The cleverness is accentuated by the discovery that it's not just the mosquitoes that the monkeys are trying to ward off with their millipede ministrations. They also suffer from the larvae of the bot fly, which the mosquitoes bring with them. When the mosquito alights to feed, the monkey's body heat triggers the hatching of the eggs, releasing larvae that burrow under the monkey's skin. They live there for several weeks, causing large, painful sores – unless the monkeys ward off the mosquitoes.

The chemicals that the capuchins have found are many times more potent that anything in human pharmacies – and they taste like it. In a detailed study of the practice, Valderrama found that many of the monkeys don't just rub the millipedes on to their fur; they will pop one or two into their mouths first and bite, presumably to make sure they release all the chemicals. Humans who have copied them say the compounds taste worse than nasty, they are very painful. And the monkeys seem to agree. After biting into the millipedes, they drool and roll their eyes in obvious pain. But they still do it. 'You'll see one or more monkeys looking frenzied and agitated, their bodies contorted as they're patting themselves all over,' says Valderrama. Then they pass the millipede on to the next in line. (But, say researchers, don't try this yourself. Benzoquinones are a known cancer agent, and while taking the chance of cancer may make sense for monkeys, we have some safer alternatives.)

There is some evidence that other primates have cracked the potential of some other plant and insect products as antibiotics, analgesics and even mind-altering hallucinogens. Sceptics point out that many plants contain such a range of chemicals that it is hard to be sure that the animals know exactly what they are doing. But, on the other hand, if they ate a potent hallucinogen and didn't like the effect, they would probably avoid it next time round. If they go back, junkies or not, they must want it.

Many researchers are bowled over by such pharmacological and tool-making skills. But they do not necessarily make apes smart, says Daniel Povinelli of the University of Louisiana at Lafayette. He famously defected from the 'smart chimp' camp after spending days watching the chumps learning to beg for food from a human trainer. They managed the basic task well enough, but then they spoilt the effect by dismally failing to spot that begging didn't deliver the goods when the trainer had a paper bag over his head.

Arguably, they were being no worse than a small child who thinks he is invisible when he covers up his own eyes. And assuredly there aren't too many paper bags in the jungle for them to be familiar with. Povinelli argued that if they were really capable of learning new things, they ought to have cracked the subterfuge. But Brian Hare of Harvard University takes a different view. 'It is easy to demonstrate that an animal is stupid. I could make Povinelli look stupid in a novel situation. The trick is to find situations in which they can demonstrate what they can do.'

* * *

A more recent way of assessing how close apes are to humans has been to go gene-counting. Headline writers are fond of pointing out that the great apes share 98 per cent of their genes with us. So, the argument goes, they must be 98 per cent human. But such sharing of genes is not too impressive, actually. Humans share half their genes with fish and a quarter with dandelions. Does that make humans one-quarter dandelion?

If genes and technology don't render humans distinct, how about social systems? After the demise of the tools theory of human distinctiveness, we have tended to define ourselves as different because we have a cultural dimension. This might be defined as engaging in activities that different groups develop for themselves in different places and at different times, and that seem to serve no special purpose other than as expressions of the group. It might be a special means of communication, or some form of artistry or even an unusual sexual activity. But here again, recent research shows that we humans are not so special. Different groups of chimps, it turns out, adopt different methods of grooming, hunting and eating. They find new ways to crack nuts and trap ants and then pass the ideas on to their fellows, who follow suit. So far as researchers can tell, these special ways of doing things are unconnected with their local environment and are unlikely to be genetically programmed in any way. It seems they just decide as a group to do things differently. That, say the researchers, is culture.

For a while, people thought that only chimps and humans did this. But recently, Carel van Schaik of Duke University in North Carolina has reported on a wide range of similar, apparently cultural behaviours among different groups of orang-utans. There are only around fifteen thousand orang-utans left in the wild, holed up in the jungles of Borneo and Sumatra – two forest islands that are losing their trees faster than probably anywhere else on Earth. They form tribes that learn and pass on to their members and the next generation distinctive tool-making skills and ways of living that look like rudimentary forms of culture. Two groups in particular have caught the researcher's eye, or rather ear. Before they retire for sleep each night, they set up a communal round of catcalling. No other groups do it; only these two groups. But one group is in northern Sumatra, while the other is in the far northeast of Borneo, in the Malaysian province of Sabah. Nobody knows why they do it; but they do.

Other tribes of orang-utans have learned to use a stick to extract seeds from fruit on the Neesia tree without getting caught by stinging hairs. In a couple of other groups the males regularly masturbate with sticks; but elsewhere the males, whether through prudery or a lack of sexual experimentation, don't indulge in the stick trick. Van Schaik has compiled a long list of these cultural activities practised by various tribes of orang-utans in Sarawak and Borneo. Many sound like the antics of human children left on their own in the woods on a camping weekend. They include making squeaking noises by blowing on to leaves or through clenched fists; building nests for play; using leaves as gloves to handle spiky fruit; pushing over dead trees; using hollow twigs to suck up ants; building nests that connect trees on either side of a river; biting through a vine to make it the right length to swing Tarzan-like across a clearing; and blowing raspberries.

Most primatologists think that the great apes of the surviving jungles could be exhibiting a kind of social sophistication and intelligence that they share with us alone. But some researchers think it may be widespread among mammals well beyond the primate world. Frans de Waal of Emory University in Atlanta forecast in the *New York Times* that 'In the coming twenty years, we will have a host of studies on cultures in all sorts of animals.' Rats, birds and even fish do it, he said. 'We will not think of culture as a monolithic thing, but a concept that includes songbirds, the great apes and humans.'

But much may depend on what we mean by culture. It is far from clear that one group of fish engage in cultural activity that other groups from the same species do not, but we shall see. Most believe that social structures among primates are far more complex and subtle than among other animals. They seem to form groupings at many different levels, from kinship groups to tribes and even up to what some term nations, with national characteristics. Several researchers have noticed how groups of chimpanzees and orang-utans that occupy the same forest, while living distinct lives and having their own distinct ways, show more similarities than other groups in distant forests.

Most interesting, whenever African chimps move from one group to another – after a dispute, say, or because the single group has become too large – the migrant swiftly adopts the habits of the new group. For many, this is seen as a clincher for the cultural status of these activities. The different hand gestures that chimps use during grooming seem to be of special importance to each group. In some groups, when two animals meet and start to groom they join their left hands above their heads. Others join their right hands. This appears to have no practical function, so it is likely to be a purely social gesture. 'It's like the boy scouts having a three-finger salute and the cub scouts having a two-finger salute,' says William McGrew of Miami University in Ohio.

He has noticed that two neighbouring groups of chimps in the Mahale Mountains National Park in Tanzania have slightly different hand gestures. One group adopts a palm-to-palm grasp, while the other has a wrist-to-wrist grip. And chimps who migrate from one community to the other change grips. 'When we see that kind of nuance in human behaviour, we call it culture,' says McGrew.

* * *

But culture has a darker side. Humans, we once believed, were the only animals to spend large amounts of time hunting and killing – not for food or self-defence or to protect territory, but purely for pleasure. At last, this seemed to be something that genuinely distinguished us from the rest of the primate pack. But even here, it transpires we are not alone. For it turns out that chimpanzees too go in for ritualized hunting for pleasure. The male chimps form hunting parties that go after many different species, such as pigs and antelopes. But their favourite – at any rate in the Gombe National Park in Tanzania, where Jane Goodall began her pioneering chimp researches forty years ago – is the chase for the red colobus monkey.

There are some unnerving aspects of these hunts that sound awfully human. It is largely a male activity, and it is social. Lone hunters are found, but mostly it is what chap chimps do together. And not necessarily chap chimps who are closely related. John Mitani, an anthropologist from the University of Michigan, says his research among a group of 140 chimps in Kibale Park in Uganda found that 'male chimpanzees cultivate social relationships not with brothers but opportunistically with others of the same age'. Like human urban male gangs, they become brothers through shedding blood, rather than being blood brothers.

A typical hunt goes as follows. A posse of male chimps go out to find a few monkeys in the trees and ambush them. They herd them towards the head hunter, who is hiding. He leaps out, causing a panic among the monkeys, during which the rest of the gang pounces in a cacophony of shrieks and hoots that echoes through the forest canopy. Afterwards the hunting party share the spoils.

Despite the eating that follows the killing, these hunts seem to be as much for fun as for food. Certainly a lot of fun is had along the way. Often the killing takes a very long time, because the chimps bite their monkey victims in a form of torture, and throw them repeatedly against tree trunks and between each other, before the serious business of killing begins. And while there is a hunting season, it is not obviously tied to when the chimps are hungry. Chimps are largely vegetarian when not hunting, but they have periods of what Goodall calls 'hunting crazes', during which they go out killing every day regardless of the need for food and apparently just for the joy of killing.

'Hunting binges occur especially when the females are sexually receptive,' says Craig Stanford of the University of Southern California. It becomes, he says, 'an orgy of meat eating and sex straight out of *Tom Jones*', in which meat is given to favoured females as a reward for sexual favours. But others disagree. Mitani says he found little evidence of that: where he has done his research, 'males share meat in order to curry favour and support of other males', he says – though that, too, may be a local cultural tradition. People such as Mitani put these hunting sprees down not to sex but to tribal politics, and especially so when the chimps do the most disturbingly human thing of all – when they go hunting each other.

These battles, like human wars, often seem to be about territory and border disputes. Chimpanzees are highly jealous of territory and patrol their borders constantly. In one widely reported study, one chimp war broke out after nine members of a tribe moved to a nearby valley to start a new group. Soon after they left, males from the first community set out on raids to hunt down the breakaway chimps. Posses of five or six males would slowly enter enemy territory and hang around until they found a chimp alone. Then they would attack it. After a raid, the hunters returned to their own area of the forest, leaving their badly injured victim to die.

In another case, a large group of chimpanzees systematically hunted down every member of a small group, killing them with their bare hands and feet. Nobody knows what triggers these border disputes in the natural environment, but there is growing evidence from Gabon that logging of the forests in the mid-1990s may have caused a series of wars. Lee White of the Wildlife Conservation Society says those animals squeezed out of their land by advancing gangs of chainsaw-wielding loggers retreat into the forest and invade the territory of their neighbours. While groups of gorillas in the forest, who are used to living together, squeezed up to make room for refugees of the logging, chimpanzees fought wars. 'This process goes on and on and on as the loggers move through,' he says. Thousands of chimps may have died.

We once laughed as zoo-keepers dressed up chimpanzees as humans and gave them tea parties. Chimps were chumps. Now, the laugh is on us. Their gentle unconscious parody now seems like the darkest satire.

* * *

But if our chimpanzee ancestors are turning out to be more gratuitously (and humanly) violent than we thought, then their closest relatives are turning out to be more gratuitously (and humanly) sexual. The bonobos, once known as pygmy chimpanzees but now recognized as a distinct species, live in the Congo basin, and they have a very marked cultural trait: sex. As Stanford puts it, 'They mate more often, in more positions and with more recreational than procreational intent, than any mammal other than Homo sapiens.' How do you tell if their sexual athletics are recreational or procreational? Well, for one thing they seem unusually fond of homosexuality, and that is hardly procreational.

Female bonobos get it together by rubbing their genital swellings together. (Primatologists say they are easing tensions between individuals, though that could easily say as much about their own sex lives as those of the bonobos.) Meanwhile, male bonobos also engage in same-sex genital rubbing – something no macho hunting chimp would be seen dead doing. Bonobos, it has to be said, like paedophilia, too. In fact, says veteran bonobo voyeur John Watkin, 'Sexual interactions occur between any combination of individuals of any sex and age.'

Bonobos make love, not war, says Frans de Waal. 'The chimpanzee resolves sexual issues with power; the bonobo resolves power issues with sex.' If chimps are from Mars then maybe bonobos are from Venus. How curious to find all this peace and love in the land of Conrad's *Heart of Darkness*.

LOSING OUR RELATIVES

Our nearest ancestors all seem close to extinction in the wild. Asia's only great apes are down to their last fifteen thousand or so, all living on Borneo and Sumatra, two islands that are losing their rainforests faster than anywhere on Earth. Two-thirds of their habitat is lost, according to the UN Environment Programme (UNEP). Probably half of the world's population disappeared during the 1990s, as their habitat was eaten up by loggers, forest fires, conversion to oil-palm plantations and fragmentation by roads and mines.

In Africa, gorillas currently number around 100,000, though subspecies like the Cross River and Mountain gorilla are numbered in their hundreds. There still may be as many as 200,000 chimpanzees, but they are thought to be disappearing so fast that they 'could be extinct within fifty years', according to a major conference on great apes held in 2004. One important chimp subspecies, confined largely to Nigeria, is likely to be gone within twenty years.

The bonobos are already down to their last twenty thousand individuals, according to some researchers, and UNEP estimates that they are disappearing at a rate of almost 3 per cent a year. These African great apes are suffering as much from hunting for the continent's large bushmeat business as they are from habitat loss.

THE GHOST OF OTA BENGA

Back in 1906, the Bronx Zoo in New York had been in business for a mere eight years. Its eccentric director William T. Hornaday had assembled a thousand or so animals in twenty-two exhibits, including his much-prized herd of American bison and a new display of snow leopards. But it was jungle apes from Africa that he wanted most and, in September that year, the crowds thronged to his latest display. In the new monkey house that summer, Hornaday offered monkeys, chimpanzees, an orang-utan named Dohung, a gorilla named Dinah – and an African pygmy called Ota Benga.

The cage became hugely popular and the crowd bayed and tried to paw this 'half-man, half-ape'. One day the zoo had a record forty thousand visitors. 'The sudden surge of interest was entirely attributable to Ota Benga,' said the *New York Times*, which published a poem declaring that he had been brought 'From his native land of darkness/To the country of the free/In the interest of science/And of broad humanity'.

Ota Benga had been brought back from Central Africa, apparently of his own free will, by an American animal collector called Samuel Verner, who bought him at a slave market on the Kasai river. He had been captured during a raid on his village by the notorious thugs, the *Force Publique*, of King Leopold of Belgium, who ruled the Congo basin as his private fiefdom. It was at the height of their reign of terror.

The unfortunate man had been persuaded to go to the US by Verner, who was collecting human exhibits for display at the St Louis World's Fair in 1904. He formed part of a tableau showing 'emblematic savages' living in a typical African village, part of 'an exhaustively scientific demonstration of the stages of human evolution'. No prizes for guessing that the forest-dwelling pygmies came at the bottom of the pile. As *Scientific American* said at the time, the Congolese pygmies were 'small, apelike furtive creatures [who] live in dense tangled forests in absolute savagery'. Verner took Ota Benga back to Africa after the fair, but when he was ostracized by his own tribe, he returned him to the US and lodged him with Hornaday, who put him back on public display.

Ota Benga, with his filed teeth and boyish manner, was caught up in a complex debate, triggered by Charles Darwin's ideas on evolution. Certain evolutionary biologists of the day spoke of a 'close analogy of the African savage to the apes'. Physiologists said Ota was 'not much taller than an orang-utan, their heads much alike, and both grin in the same way when pleased'. Psychologists claimed that pygmies 'behaved a good deal in the same way as the mentally deficient person'. 'Was he a man or a monkey?' the *New York Times* asked, before answering that while 'pygmies are fairly efficient in their native forests, they are very low in the human scale'.

Hornaday 'saw no difference between a wild beast and the little Black man'. Verner, himself a trained academic, said that pygmies were 'the most primitive race of mankind . . . almost as much at home in the trees as the monkeys'. In their homeland he wanted them collected in reservations so that their lands could be colonized by 'the white race'. Indeed when black clergy in New York sought to rescue Ota Benga, claiming that 'our race is depressed enough without exhibiting one of us with apes', *The Times* responded imperiously that this was an affront to science and 'the reverend coloured brother should be told that evolution is now taught in the textbooks of all schools'.

The zoo finally relented and removed Ota Benga from display. He spent some years under the protection of clergy in a seminary in Vermont, but in the autumn of 1916 – finally realizing that he would never return home – he went into the woods and shot himself in the heart. A very American end. To this day Ota Benga remains used: the Bronx Museum web site has airbrushed him from its history, while creationists use his case as a weapon in their continuing war with evolutionary science, which they equate with racism.

* * *

The pygmy people of Central Africa, like no other people, seem to represent our forest past. Their residence in the African forest probably goes back tens of thousands of years, whereas the Bantu people – the dominant peoples of most of sub-Saharan Africa – are much more recent arrivals, and much less at home in the forests. According to experts like Douglas Wallace of Emory University in Atlanta, recent detailed analysis of the DNA of the world's peoples shows that the pygmies are probably most genetically similar to the first

Homo sapiens, who roamed East Africa some fifty thousand years ago – the first people of our own species, and the people from whom every other race and group around the planet evolved. 'We are looking at the beginning of what we would call *Homo sapiens*,' Wallace said recently of the pygmy people.

This is not to denigrate them. Far from it. They, of all people, are still able to show us the rainforests as our ancestors hundreds of generations ago would have known and understood it. But their reign is passing. Only a few of the pygmy tribes can still survive untroubled, alone in the forest, living by hunting and gathering rather than farming or trade. In Congo-Brazzaville the Bayaka people are the most famous of these, able to track and kill elephants with spears, call antelopes to them, charm gorillas from the trees, identify the hundreds of different snakes and thousands of medicinal plants, and guide Western scientists safely through the great jungles of the Congo basin.

Often subjugated by their Bantu neighbours when living in villages, they will spend months or years in the jungle. And yet, legally, many have been put beyond the law by governmental bans on hunting for bushmeat in the national parks of the region. The Bayaka, for instance, are forbidden to hunt in their traditional domain, even with their age-old implements of spears and home-made crossbows. They know the forest like the backs of their hands, and they hunt only what they need for the pot; and yet, because others have hunted down the forest animals with automatic weapons to fill supermarket shelves in cities, they too are prevented from hunting.

They may often quietly ignore this rule. Who wouldn't? And conservationists have tried to preserve the Bayaka's skills, livelihoods, and self-esteem by utilizing their tracking abilities to find animals for research rather than hunting. But the risk is that parks set up to save gorillas may cause the extinction of the ancient ways of their human inhabitants, the Bayaka. Whatever our environmental or even ethical concerns, can it be right that these people, who are among the last hunter-gatherers left on the Earth, our ancient ancestors should have their traditions outlawed in this way? Ota Benga, looking down from a jungle in the sky, might well take the view that little has changed since he put a gun to his heart in a Vermont woodland ten thousand kilometres from his rainforest home.

JUNGLE INDISCRETIONS

We still allow things to go on in the jungle that would not be tolerated elsewhere – whether it is cultural denial in the Congo, the genocide in Rwanda or the brutalizing of children as soldiers in West Africa. The jungle, literally and metaphorically, still cloaks dark deeds. And, to this day, we still cannot work out what to make of the people of the forest. Like the forest itself, at times they are hopelessly romanticized as 'noble savages' living in some timeless Eden and in possession of magical gifts.

At other times they are demonized or denigrated as inferior beings. Take the long dispute over the body of an African bushman, reportedly a tribal chief, stolen from his grave in Botswana shortly after his death in the 1830s by two French adventurer-naturalists, Jules and Edouard Verreaux, who took the body to France, stuffed it and sold it to a naturalist from Barcelona who was building a collection of African artefacts. The collection was eventually amalgamated with that of the Museum of Natural History at nearby Banyols, where the stuffed bushman, known as '*El Negro de Banyols*', went on display in 1916.

And there it stayed behind glass until 1992, when the city fathers temporarily removed it from display, fearing that its presence on display could lead to an African boycott of the 1992 Olympics in Barcelona. It was finally removed 'out of respect to the thousands of African immigrants who live in the town'. One night in September 2000, the corpse was secretly removed from the museum and sent back to Botswana, where it was buried in an official ceremony.

LANGUAGE AND DIVERSITY

Perhaps humanity's greatest intellectual and practical accomplishment is language. It is language above all that separates us from our nearest relatives. And where do we see it in its greatest complexity? Most of the world's languages are to be found in the rainforests. There are an estimated 6,700 languages worldwide, according to the UK-based Foundation for Endangered Languages. Occasionally new ones are discovered in linguistic heartlands such as Papua New Guinea or Indonesia. Indeed these two countries, home of most of the remaining rainforest of southeast Asia, account for a quarter of all the known languages in the world – a linguistic equivalent of the rainforest biodiversity 'hotspots'. Other hotspots are in the Congo and Amazon basins. Many of these languages are spoken by only a few people. The Gurumulum speakers in the Papua New Guinea rainforest number just ten.

There is a near-exact correlation between species diversity in the natural world and language diversity among humans. Both are centred on rainforest regions. Tropical rainforests, which cover 6 per cent of the land surface and contain at least half of the world's species, also contain approaching half of the world's languages. You can see why. Rainforest dwellers live self-contained lives in small tribes. They don't get out a lot. They don't need to. More particularly, people living close to nature, modifying and adapting to it, develop an often very specialized knowledge about their environment. Their knowledge is very local, and so, therefore, is their language.

And just as species disappear, so do languages. Around half of the world's languages are reckoned to be moribund, in that they are not being passed on to the next generation. The demise of some can be dated precisely, says linguist Bruce Connel in the Foundation's newsletter: 'Kasabe in the Adawawa region of Cameroon had one speaker, a man called Bogon. On 5 November 1995 he died, taking Kasabe with him. His sister can understand the language, but does not speak it. His children and grandchildren do not know the language.' As each language dies, science – in linguistics, anthropology, prehistory and psychology – loses one more precious source of data, one more of the diverse and unique ways that the human mind can express itself through a language's structure and vocabulary.

HUMAN IDENTITY IN THE FOREST

It is time to dispense with the myths. Green heaven and green hell; jungle fit only for clearing and rainforests as primeval wilderness that must be preserved at all costs; forest dwellers as noble savages or as half-human head-hunters and witch doctors with their pots. They all have to go. The truth is different. More complicated, more nuanced, more ambiguous and more interesting. Many, perhaps most, rainforests are partly human constructs. In recent millennia, there have rarely been wildernesses untouched by humans: they have been cleared and culled and planted and abandoned. They are, rather, wild gardens created when people in the jungle had greater wisdom about how to manage them than Westerners, at any rate, seem to have today.

That in no way devalues the forests. As we have seen, they remain the depository of most of the biological wealth of the planet, in both volume and diversity. They are hugely important for the maintenance of our atmosphere, our climate, our rainfall. They are vital for the future habitability of our planet. The more we learn, the more we realize how important they are to life on Earth. But it does make them part of us, not something other. Billions of people live in or close to the rainforests. Hundreds of millions live in those areas that biologists have deemed to be biodiversity 'hotspots'. Even if we wanted to, we cannot simply erect fences around these wildernesses and leave them to nature. In most places, we have to learn to use and preserve them at one and the same time. And the discovery that our ancestors did just this should encourage us to try afresh.

We came from the jungle. We became human after stepping down from the trees, probably during one of the periodic eras when the rainforests of Central Africa dried and became grasslands. We learnt then to hunt on the ground rather than in the forest canopy, to walk upright rather than swinging from the branches, and to use our hands instead to make tools, to control fire and to clear the forest for farming. Perhaps that departure has become a symbol of our divorce from nature. Perhaps it explains some of our confusion about the forests and our constant mythologizing about them.

But for millennia we lived in plausible harmony with the jungle. We cleared it, but it recovered, time and again. Now it is time to resurrect that connection. For our own future as a species we need to stop seeing jungles as alien and other. As human beings in the twenty-first century, many of us long to find our personal roots, our ancestral origins. Now surely it is time that we rediscovered our roots as a species and made peace with our past.

FURTHER READING

INTRODUCTION

Grove, R.H. (1995). *Green Imperialism: Colonial Expansion, Tropical Island Edens and the Origins of Environmentalism, 1600–1860*. Cambridge: Cambridge University Press.

Slater, C. (2002). *Entangled Edens: Visions of the Amazons*. Berkeley: University of California Press.

CHAPTER 1

Cocker, M. (1998). *Rivers of Blood, Rivers of Gold: Europe's Conflict with Tribal Peoples*. London: Pimlico.

Coe, S.D. & Coe, M.D. (1996). *The True History of Chocolate*. London: Thames & Hudson.

Desmond, R. (1995). *History of Kew*. London: The Harvill Press.

Drayton, R. (2000). *Nature's Government: Science, British Imperialism and the Improvement of the World*. New Haven: Yale University Press.

Ecott, T. (2004). *Vanilla: Travels in Search of the Luscious Substance*. London: Michael Joseph.

Forbath, P. (1978). *The River Congo: The Discovery, Exploration and Exploitation of the World's Most Dramatic River*. London: Secker & Warburg.

Hemmings, J. (1970). *The Conquest of the Incas*. London: Macmillan.

Idriess, I.L. (1933). *Gold Dust and Ashes: The Romantic Story of the New Guinea Goldfields.* Melbourne: Angus & Robertson.

Jardine, L. (1999). *Ingenious Pursuits: Building the Scientific Revolution.* London: Abacus.

Reynolds, E. (1985). *Stand the Storm: A History of the Atlantic Slave Trade.* London: Allison & Busby.

Severin, T. (1997). *The Spice Islands Voyage.* London: Little, Brown & Co.

Shoumatoff, A. (1979). *Rivers Amazon.* London: Heinemann.

CHAPTER 2

Conrad, J. (1973). *Heart of Darkness.* London: Penguin.

Duran-Reynals, M.L. (1946). *Fever Bark Tree.* London: W.H. Allen.

Fuller, E. (1995). *The Lost Birds of Paradise.* Shrewsbury: Swan Hill Press.

Griffiths, T. & Robin, L. (eds) (1997). *Ecology and Empire: The Environmental History of Settler Societies.* Edinburgh: Keele University Press.

Hecht, S. & Cockburn, A. (1989). *The Fate of the Forest: Developers, Destroyers and Defenders of the Amazon.* London: Verso.

Hochschild, A. (1998). *King Leopold's Ghost: A Story of Greed, Terror and Heroism in the Congo.* Boston: Houghton Mifflin.

CHAPTER 3

Baker, O. (2002, 9 February). 'Law of the Jungle'. *New Scientist,* p.28.

Caufield, C. (1985). *In the Rainforest.* London: Heinemann.

Hubble, S.P. (2001). *The Unified Neutral Theory of Biodiversity and Biogeography.* Princeton: Princeton University Press.

Rabinowitz, A. (2001). *Beyond the Last Village: A Journey of Discovery in Asia's Forbidden Wilderness.* New York: Island Press.

Wilson, E.O. (1992). *The Diversity of Life.* Boston: Belknap Press.

CHAPTER 4

Adams, J.A. & McShane, T.O. (1996). *The Myth of Wild Africa: Conservation without Illusion.* Berkeley: University of California Press.

Fairhead, J. & Leach, M. (1998). *Reframing Deforestation: Global Analyses and Local Realities – Studies in West Africa.* London: Taylor & Francis.

Litvin, D. (2003). *Empires of Profit: Commerce, Conquest and Corporate Responsibility.* New York: Texere.

Yuscaran, G. (1995). *Gringos in Honduras.* Tegucigalpa: Nuevo Sol.

CHAPTER 5

Collins, M. (ed). (1990). *The Last Rain Forests.* London: Mitchell Beazley.

Garrett, L. (1994). *The Coming Plague: Newly Emerging Diseases in a World out of Balance.* New York: Farrar Straus and Giroux.

CHAPTER 6

Colchester, M. & Lohmann, L. (eds). (1993). *The Struggle for Land and the Fate of the Forests.* London: Zed Books.

Monbiot, G. (1991). *Amazon Watershed.* London: Abacus.

GENERAL

Hansen, E. (1988). *Stranger in the Forest: On Foot Across Borneo.* London: Century Hutchinson.

Morris, J. (1973). *Heaven's Command: An Imperial Progress.* London: Faber & Faber.

Reader, J. (1997). *Africa: A Biography of the Continent.* London: Hamish Hamilton.

INDEX

Acosta, Jose de, 12
agoutis, 120, 123–5
AIDS (HIV), 2, 163, 197, 235, 236–7, 246
Akowuah, Daniel Kwaku, 189
Alexander, John, 160
alligators, 36
Altshuler, Douglas, 121
Amazon: biodiversity, 177–8, 260, 302; early settlement, 170–2, 172–6, 179–81; European exploration, 14–15, 38, 41, 48–50, 99–100; foreign settlers, 277–9; languages, 302; manatees, 228; rainforest, 110–11; rubber, 77–8
Amazons, 13–14, 18, 175
Angel, Jimmy, 24
Angkor, 165–7, 191
ants, 126, 138–9
apes: bushmeat, 216, 234; cultural activity, 288–90; diseases, 234; genes, 287; hunting, 291–3; intelligence, 286–7; numbers, 295; sex, 294; tool-using, 283–5; warfare, 293
Arana, Julio Cesar, 78, 93, 139
Aruak people, 16
Ashanti people, 27
ayahuasca, 45
Ayres, José Marcio, 228–9
Aztecs, 11, 12, 29–30, 74, 99
Balee, William, 177–8
bananas, 182, 183, 192–5
Banks, Joseph, 39
Bantu people, 192, 299
Bänziger, Hans, 127–8
Barro Colorado Island, 115–19, 120–1, 144
Bates, Henry, 41, 47–50
bats, 121, 232–3
Baures, 170–2
Bayaka people, 141, 299–300
Bayliss-Smith, Tim, 183
Beebe, William, 143
beef, 212
bees, 7, 30, 124
beetles, 107, 112, 125–126
Ben-Amos, Dan, 158
Ben and Jerry's ice-cream, 129, 255
biodiversity, 112–15, 142, 177–8, 210, 244, 260
bioprospecting, 241–57
birds of paradise, 4, 51, 59, 80–2
Blake, Steve, 130
Blood, Captain Ned, 82
boa constrictors, 37
Body Shop, 255
Bolivia: border, 23, 78; cinchona, 69–71; early settlement, 170–2; farming, 263, 264–5; peanuts, 72; rubber, 77–8
bonobos, 294–5
Bonpland, Aimé-Jacques-Alexandre, 35, 37
Borneo: canopy crane, 144; elephants, 131–2; ethnobotany, 250; gold fraud, 24–5; logging, 199–202, 204, 295; orang-utans, 204, 288–9, 295; peat-swamps, 111, 201, 206; rainforest, 111; Wallace Line, 59
Bostwick, Kim, 126
Botswana, 301
Brazil: biodiversity, 112, 113, 210; deforestation, 212–13; drugs companies, 246–7; manatees, 228; rainforest, 38, 204, 210; rubber, 76–9, 95–7, 270–3
Brazil nuts, 37, 119, 122–5, 129, 146, 248, 270
Bre-X, 24–5
Bubb, Philip, 103
Burma, 40, 110
Burroughs, William, 100, 101
Busang, 25
bushmeat, 197, 214–26, 234, 237, 258, 295
butterflies, 44, 49, 112

cacao, 29
caffeine drinks, 45
Calabar bean, 33
Calancha, Antonio de, 62
Cambodia, 166
Cameron, Verney Lovett, 87–8, 89, 91–2, 161

Cameroon: bushmeat, 219, 220, 222, 224; cocoa, 266–70; diseases, 237; early settlement, 164; Korup Park, 261–2
Cao, Diogo, 27–8
capybaras, 36
carbon dioxide, 147–8
Carvajal, Gaspar de, 13, 14
Casement, Roger, 93
Catherwood, Frederick, 168
cats, 133
Charles II, King, 32, 63
Chibcha people, 12, 14
Chileno, Andreas, 263–4, 285
chimpanzees, 284–5, 290–4, 295
China, 40, 204, 209, 227
Chinko river, 258
chocolate, 269; drink, 29–30, 32
Choe, Jae, 120–1
cinnamon, 29, 43
cinchona, 61–71, 247, 249
City Z, 24
Clapperton, Hugh, 230
Clark, David and Deborah, 169
cloud forests, 102–3, 111
cloves, 42–3
Clutton-Brock, Juliet, 159
coca, 30–2
cock of the rock, 50
cocoa, 266–70
coffee, 40
Colombia, 78, 99, 113
Columbus, Christopher, 11–12, 27, 29, 71
Commerson, Philibert, 55
Conan Doyle, Arthur, 23, 24, 137
Condamine, Marie de la, 35, 37, 74
Congo: biodiversity, 210, 302; bushmeat, 222–4; early settlement, 163, 175; European exploration, 87–9; ivory trade, 90, 91–2, 94; jungle, 28, 162–3; Leopold's kingdom, 89–94; logging, 211–12; Mokele Mbembe stories, 141; Prester John myth, 27; rainforest, 110, 211; river, 27, 88, 230; rubber, 92–3, 94
Conrad, Joseph, 89, 93–4, 294
Conservation International, 136
Cortez, Hernando, 12, 29, 71, 74
Cracraft, Joel, 133–4
Cromwell, Oliver, 63
Cross, Robert, 96
cryptozoology, 141
cunani bush, 243–4
curare, 35, 37, 247

Darling, Patrick, 153–61
Darwin, Charles: Beagle expedition, 38, 51; on Brazilian rainforest, 38, 112; evolution theory, 4, 41, 51–3, 56, 107, 281, 297; illness, 41, 51, 230; orchid moth prediction, 56–57; Wallace relationship, 48, 59, 51–3
Dayak people, 205
de Waal, Frans, 289, 294
deer, 135–6, 227
deforestation, 184–91, 232
Dial, Roman, 144
diseases: in Africa, 231; Chagas, 51, 230; cures, 34, 55, 246–8; effects of European diseases in South America, 21; mosquito-borne, 231–2; new, 232–7; in South America, 15–16, 230; syphilis, 21; yellow fever, 231
Dortal, Jeronimo, 16–17
Dunlop, John, 78–9, 97

Ebola virus, 163, 197, 233–5
Ecuador, 44, 78, 101–5, 184
El Dorado, 4–5, 9, 11, 14–22, 29, 36, 175–6
electric eels, 37
elephants, 90–2, 128, 130–2, 214, 221, 258
Emmons, Louise, 136–7
Erickson, Clark, 170–2, 176, 177–8
Essessima, Joseph, 266, 268
ethnobotany: benefits to local people, 246–8; commercial value, 252–7; drugs companies, 246; early work, 100; effects on habitat, 249–50; indigenous attitudes to, 242–5; usefulness, 250–1
Fa, John, 216–22, 223, 225
Fawcett, Colonel, 22–4
Fay, Mike, 161–2
Flynn, Errol, 22, 81
Ford, Henry, 277
Frederick, Adolphus, 1
Freud, Sigmund, 31
frogs, 125, 251
fruit, 182
Fusani, Leonida, 127

Gabon, 274–6
Galapagos Islands, 41, 104
Galton, Francis, 85
Gama, Vasco da, 28
Gemerden, Barend van, 164
Ghana, 27, 97, 188–9, 217
Glaser, Bruno, 180
globalization, 212
Gockowski, Jim, 269
Godoy, Ricardo, 252, 253–7
gold: in Africa, 27–8, 39–40; in Americas, 12, 14–21, 28; fraud, 24–5; in New Guinea, 22; search for, 11–13, 15–21, 27
Gomes, Fernao, 27
Goodall, Jane, 283–4, 292
Goodyear, Charles, 75–6
gorillas, 216, 234, 295
Gorinsky, Conrad, 242–5
greenheart tree, 243
Guaja people, 177
guarana, 45
Guatemala, 167–8
Guinea, 187
Guzman, Michael de, 25

Haffer, Jurgen, 113
hallucinogens, 29–30, 45, 98–101, 286
Hayse, Bruce, 258–60
Heckenberger, Michael, 173–4
Hemming, John, 16, 18
Henry the Navigator, 9, 27
Hernandez, Francisco, 30
Herzog, Werner, 2
hippos, 221–2
HIV see AIDS
Holden, Jeremy, 132
Honduras, 167, 254–7
Hooker, Sir William, 45–6, 95
Hooker, Joseph, 95–6
Hornaday, William T., 296–8

Hubbell, Steve, 115–19, 120
Humboldt, Alexander von: commercial views, 46, 64; curare investigation, 35, 38; explorations, 36–7, 129; ideal of jungle society, 2; influence, 48; opinion of Amazon Indians, 38, 179
hummingbirds, 37
Hutten, Philip von, 19
Huxley, Aldous, 100

Indonesia: economy, 202–3; forest fires, 114; logging, 202–4, 207; palm-oil, 213; peat-swamps, 111, 206–7; rainforest, 110
Ingram, Sir William, 82
Irrawaddy river, 110
ivory, 90, 91–2, 94, 257
ivory-nut palm, 40

Japanese farmers, 277–8
Jesuits, 61–3, 172
Jost, Lou, 101–2, 103–5

Kabila, President Joseph, 211
Kaimowitz, David, 212
Kalimantan, 201, 203–5, 226
Kayapo people, 129, 181–2
Kealhofer, Lisa, 166
Keith, Minor Cooper, 195
Kew Gardens, 39–40, 45–6, 67–70, 95–6
Kingsley, Mary, 231
Korup project, 261–2
Kuikuro, 17

La Selva, Costa Rica, 169
languages, 302–3
Leach, Melissa, 186–7
Leary, Timothy, 101
Ledger, Charles, 68–70
Lee, Kenneth, 170, 172
lemurs, 110
Leopold II, King of the Belgians, 88, 89–91, 164, 296
Lesson, Rene, 81
liverworts, 46, 105
Livingstone, David, 1, 47, 84–6, 87–8, 230
Lloyd, Peter, 156, 160
logging, 199–202, 211
McConnell, William, 184–6
McGrew, William, 290
Macintosh, Charles, 75
McKinnon, John, 135–6
McNeely, Jeff, 262
Madagascar: biodiversity, 260; ecology, 55, 109; forest cover, 184–5; orchids, 56–7; ordeal poisons, 33; rosy periwinkle, 55, 246, 248
Magellan, Ferdinand, 42, 80
malaria, 61, 63, 84, 231, 232
Malaysia, 97, 110, 112, 131, 213
Mamani, Manuel Incra, 68–9, 70, 71
manakin birds, 126–7
manatees, 228
mangroves, 111
Marajo, 172–3
Marburg virus, 235–6
Markham, Clements, 66–8, 70, 95, 248–9
marmosets, 137
Martin, Claude, 229
Mauritius, 43
Mayan civilization, 151, 167–9, 190, 191, 195
Mazatec people, 98
medicines, 29, 33, 100, 227, 243–8
Meggers, Betty, 179–80
Mekong river, 110
Mendes, Chico, 271–2, 273
Mexico, 11, 12, 15, 21, 29, 30
Milner-Gulland, Eleanor, 218, 221, 224
Mitani, John, 291, 292
mites, 126
Mokele Mbembe, 141
Moluccas islands, 42, 59
monkeys: bushmeat, 214, 221, 222; diseases, 237; hunted by chimps, 291–2; Korup park, 261–2; mosquito repellent, 284–5; sanctuary in sacred grove, 189; tool-using, 283–4
Montezuma, 29
Morel, Edmund, 93
mosquitoes, 231–2, 284–5
moths, 44, 56, 121, 126, 127–8
muntjacs, 135
mushrooms, hallucinogenic, 30, 98–9, 101
Myers, Norman, 187, 260
Nadkarni, Nalini, 145
Nason, John, 121
New Guinea: birds of paradise, 4, 81; cloud forests, 103, 111; early farming, 183; gold, 20–1, 22; mangroves, 111; rainforest, 110; snakes, 138
Nigeria, 153–60, 219–20
nutmeg, 42–3

Oates, John, 163
oil, 274–6
oil palm, 40, 134, 202, 213
orang-pendek, 141
orang-utans, 125, 197, 204, 205, 288–9
orchids, 56–7, 102–4, 124, 126
Ordas, Diego de, 15–16
Orellana, Francisco de, 13, 14, 17–18, 19, 175
Orinoco river, 14, 17, 19, 37, 110, 144
Ortiz, Enrique, 119, 122–5
Oslisly, Richard, 162–3
Ota Benga, 296–7, 298
ox, cinnamon-coloured, 135

palm oil, 202, 213
Panama, 112, 115, 117–19
Pangea, 109
Park, William 'Sharkeye', 22
Payne, Katy, 131
peanuts, 72–3
peat-swamps, 111, 201, 202, 206–207
Pereira, Pacheco, 156
Peru: biodiversity, 112, 113; coca, 31; explorations, 44, 74; rubber, 74, 78; Spanish conquest, 11, 12; value of rainforest, 253
Peters, Charles, 253, 254
Petit-Thouars, Aubert de, 56
peyote, 98
pineapples, 32, 40
Piperno, Dolores, 184
Pizarro, Gonzalo, 17, 19
poisons, 33–5, 45, 99
Poivre, Pierre, 43, 92
Posey, Darrell, 181–2
Povinelli, Daniel, 286–7
Prance, Ghillian, 113
Prester, John, 9, 26, 28
Priestley, Joseph, 74

Prunus Africana, 249–50
Purdie, William, 40
pygmy people, 192, 297, 298–9

quinine, 64–5

rabbits, 136
Rabinovitz, Alan, 136, 137
rain, 147–9
rainforests: biodiversity, 111–19; canopy, 142–6; disappearance, 186–7, 190, 208–11; distribution, 110–11; exploration, 37–8; gardening, 178–84; purposes, 239; rainmaking, 147–9; soils, 179–81, 264; types, 110–11; urban cultures, 190–1; value, 252–7
Raleigh, Walter, 19–20, 46
rats, 136–7, 226
Reede, Hendrik van, 42–3
refugia, 113
Ribadeneyra, Marcello de, 165
Richards, Paul, 112, 179
Rieley, Jack, 200–1, 206
Roosevelt, Anna, 172–3, 279
Roosevelt, Theodore, 2, 87
Roosmalen, Marc van, 137
rosy periwinkle, 55, 246, 248
Rothschild, Lionel Walter, 82
Rousseau, Henri, 2
Royle, John Forbes, 66
rubber, 71, 74–9, 92–3, 95–7, 270–3
Ruf, Francois, 266–7
Rumphius, Georgius, 43

Saavedra, Alvara de, 20–1
SARS epidemic, 236
Schaik, Carel van, 288
Schultes, Richard, 98–101, 241, 247
Seymour, Sarah, 214, 218–19, 220
Shuar people, 105
Sierra Leone, 187, 208
Simian Foamy virus, 237
slash-and-burn farming, 169, 180, 186, 264, 265
slavery, 78, 83–4, 164
Sloane, Hans, 32, 34
Smithsonian Institution, 115, 120–1, 122, 136, 179
snakes, 36, 138, 226
snuff, hallucinogenic, 45
soil, black, 179–81, 264
Solomon Islands, 110, 138, 183
spectacle bears, 105
spices, 42–3, 80
Spruce, Richard: career, 41, 44–7; illness, 41, 230; influence, 98, 100, 105; meeting with Wallace, 50; plant collection, 41, 67
Stanford, Craig, 292, 294
Stanley, Henry Morton: career, 85–6; expeditions, 47, 86–7, 88–90, 161; on ivory trade, 90–1, 92; Livingstone search, 86, 87; opinion of Africa, 86, 87; opinion of jungle, 1
Stevens, John Lloyd, 168
strangler fig, 121, 146
Suharto, President, 202, 207
Sulawesi, 202, 204, 226
Sumatra: destruction of rainforest, 204–5, 213; elephants, 132; orang-pendek, 141; orang-utans, 288, 296; peat-swamps, 111; tigers, 133–5; turtles, 226
Sungbo's Eredo, 153–61, 191

Talbor, Robert, 63
Tanghinia trees, 33
tapirs, 36, 105, 132–3
Tawahka people, 254–5, 257
tea, 40
Telfair, Charles, 193
Thailand, 40, 127, 166–7
tigers, 111, 133–5, 197, 204
Timmins, Rob, 136
trees, 117, 187–8, 278
Tuckey, James, 230
turtles, 226
Twain, Mark, 77
Tyler, Michael, 251

Unification Church, 279
United Fruit, 195

Valderrama, Ximena, 285–6
vanilla, 29, 30
Venezuela, 24, 244–5, 284
Verner, Samuel, 296–7

Waika people, 45, 46
Waldes, Nichola, 199–200
Wallace, Alfred Russel: birds of paradise, 81; career, 47–8; evolution theory, 41, 51–3; explorations, 48–51; illness, 230; Line, 54–5
Wapishana people, 242, 243–4, 245
wasps, 121, 182
water hyacinth, 139–40
Waterton, Charles, 35
Weddell, Algernon, 67
Wickham, Henry, 95–6
Wildlife Conservation Society, 130, 136, 228, 234, 293
Willis, Kathy, 163, 183
Wilson, Edward, 2, 119, 142
Wilson, James, 85
World Wildlife Fund (WWF), 135, 220, 226, 228–9, 259, 261
Wunder, Sven, 274, 275

Xinguano people, 173–4

yellow rain, 6–7
Yoruba people, 155

Zanzibar, 43

ACKNOWLEDGEMENTS

I would like to thank Bill O'Neill and the *Guardian* newspaper for commissioning a trip to Indonesia; Anne Moorhead, formerly of the International Institute of Tropical Agriculture, and colleagues for an invitation to West Africa; Xavier Viteri and Lou Jost for guidance through the cloud forests of Ecuador; John Fa and Sarah Seymour for access to their interviews with African bushmeat hunters and Patrick Darling for his insights into Sungbo's Eredo. Thanks also for help along the way to Clark Erickson, Errol Fuller, Conrad Gorinsky, Herbie Girardet, Tony Simons, Guo Yinfeng, Rob Parry-Jones, Stephan Weise, John Reader, Claude Martin, Peter Jackson, James Lovelock, David Kaimowitz, Jack Rieley, Suwido Limin, Nicola Waldes, Paul Sarfo-Mensah and Emil Frison. Also to Susanna Wadeson, Sarah Emsley and colleagues at Transworld, Mike Petty at the Eden Project, Irene Lyford for copy-editing and Jessica Woollard at Toby Eady Associates. And finally to Brian Leith, David Allen and the rest of the production team from Granada Wild whose superb TV series, *Deep Jungle*, formed the starting point and inspiration for this book, as well as providing a significant amount of the research material.